ECONOMIC AND SOCIAL COMMISSION
FOR ASIA AND THE PACIFIC
Bangkok, Thailand

ENERGY, ENVIRONMENT AND SUSTAINABLE DEVELOPMENT

ENERGY RESOURCES DEVELOPMENT SERIES
NO. 34

UNITED NATIONS
New York, 1993

ST/ESCAP/1357

UNITED NATIONS PUBLICATION

Sales No.: E.94.II.F.27

ISBN 92-1-119639-6

ISSN 0252-4368

PREFACE

The purpose of the Energy Resources Development Series is to document and disseminate the most significant development trends in the regional energy scene. In this sense the planned schedule for the publication of this particular issue was to coincide with an auspicious event in the history of ESCAP: the convening of the first session of the newly established Committee on Environment and Sustainable Development. It was only natural therefore that the documents prepared for the Committee would constitute the major part of the compilation for this publication.

In April 1992, after two years of intensive discussions and deliberations both in the sessions of the Commission and within the secretariat, the Commission decided to embark on a thematic programming approach, in contrast to the sectoral approach which had been the practice since the founding of ECAFE, the precursor of ESCAP. The 15 sectoral subprogrammes in the biennium 1991-1992 were replaced by six "thematic" subprogrammes for the biennium 1994-1995, with only transport and communications, and statistics, retaining their sectoral character. All the other sectors were abandoned in favour of three thematic subprogrammes: (1) regional economic cooperation, (2) environment and sustainable development, and (3) poverty alleviation through economic growth and social development. The subsidiary structure of the Commission was also modified and three thematic committees were established to replace the former "sectoral" committees. Thus, the first session of the Committee on Environment and Sustainable Development was held from 4 to 8 October 1993, and was preceded by the Expert Group Meeting Preparatory to the First Session of the Committee on Environment and Sustainable Development, which was held from 30 September to 2 October 1993. In the meantime the first sessions of the other thematic committees had already been convened and therefore by that time the ESCAP thematic programming approach was in full swing.

This development was radical in the sense that it represented a big departure from the customary sectoral approach. Indeed, the organizational structure of Governments in the region, as also the structure within the ESCAP secretariat, have remained largely sectoral. The divisions within the secretariat will now be obliged to service more than one committee, as their respective work programmes fall under the competence of more than one thematic committee. This new approach may be attributed to developments within the ESCAP region as a reaction to global concerns initiated by Our Common Future, the publication produced by the World Commission on Environment and Development in 1987. In 1990, The Economic and Social Commission for Asia and the Pacific called on the Executive Secretary to convene a meeting of eminent experts to develop a new strategy for the region to address these new concerns. A report was submitted to the Commission in 1991; the Commission then decided to deliberate further on the proposals set forth in the report of the group of eminent persons in a series of meetings of senior officials. A consensus was finally reached and this led to the landmark decisions of 1992. In the meantime the secretariat had prepared a revision of the medium-term plan for the period 1992-1997, which was endorsed by the Commission at its forty-eighth session in 1992. The programme of work, 1994-1995 including the subprogramme on environment and sustainable development, was endorsed by the Commission at its forty-ninth session in 1993.

The subprogramme on environment and sustainable development consists of five subthemes, one of which is energy development and management. Under this subtheme, the question of sustainable development may conveniently be classified into three "levels" corresponding to the occurrences of the energy – environment interface: global, regional or subregional, and national or local.

Thus, the ESCAP subprogramme on environment and sustainable development under the subtheme "Energy development and management" was prepared by making a judicious choice of activities among areas related to energy and electricity policy analysis and planning, including incorporation of environmental considerations, promotion of energy and electricity efficiency and energy and electricity conservation, and promotion of new and renewable energy technologies and rural energy planning. The work programme for the biennium 1994-1995 was prepared on this basis.

The question remains, what constitutes energy development and management that or is in support of sustainable development ? For the time being, we have the guidance of Agenda 21, the global programme to be pursued in future years, which was agreed upon by world leaders at the United Nations Conference on Environment and Development held in Brazil in June 1992. On matters related to energy development and management, the programme has several components under various chapter headings. In the meantime, the "no regrets policies" may be identified as being in support of sustainable development; but even then, can these be sustainable ? It is not the purpose of the present volume to seek the answer to this question, but rather only to try to look at some possibilities that may lead to some sort of managed development. A few years hence, the more discernible among these might have been investigated further and, it is hoped more positive conclusions might be drawn.

CONTENTS

CHAPTER I

ENERGY SCENE, ISSUES AND POLICIES IN THE ASIAN AND PACIFIC REGION[*]

Introduction

Energy, a necessary input for development,[1] is generated primarily from finite resources and should, therefore, be consumed at a minimal rate. However, spurred by population expansion and improved technology, the global demand for energy is ever-increasing. Most of this energy (over 70 per cent of the commercial energy supply) is generated from fossil fuels, the consumption of which is considered the major source of man-made pollution.

Fossil fuel consumption results in emissions of (a) carbon dioxide, one of the main greenhouse gases that has raised concern about climatic change or global warming; (b) sulphur and nitrogen oxides, which may cause acid rain; and (c) particulates and gases from power plants, industries and vehicles. It also creates problems of waste disposal.

Industrialized countries, which have high consumption of fossil fuels, are largely responsible for the accumulation of carbon dioxide in the atmosphere. However, as developing economies are striving to attain a high level of development, they also require a large energy input, making it likely that incremental emissions will come increasingly from the developing economies in the future. It is therefore advisable for all concerned to adopt a strategy to address the environmental issue of energy in a concerted way.

The present note covers the results of secretariat's analysis of the current energy situation in the ESCAP region in respect of resources, production, consumption, and demand. It puts forward some aspects of energy policies and issues in the area of energy developments sustainability and sustainable pathways to development. The note also discusses ways to mitigate environmental problems at the global, regional and national levels, for example, through energy conservation, increasing energy efficiency and fuel switching from fossil fuels to environmentally more benign fuels. Some regional cooperative efforts, are proposed, such as the establishment of a regional forum to discuss energy strategies and policies, an Asian energy efficiency programme for the twenty-first century, a TCDC (technical cooperation among developing countries) programme on new and renewable sources of energy and studies on regional collaboration on energy and power systems, including all traditional supply sources (oil, gas, coal and others) based on an environmentally sustainable pattern of use. The analyses are based on research, utilizing the latest data available from the United Nations Statistical Office, *BP Statistical Review of World Energy* and other publications and reports.

The secretariat's analysis contains information on the present trends and future perspectives for energy in the countries of the ESCAP region excluding the newly independent Asian republics.

I. ENERGY RESOURCES IN THE ASIAN AND THE PACIFIC REGION

Proven reserves of fossil fuels as of the end of 1992 are shown in table 1. In terms of world reserve/production ratios coal reserves continued to be the highest (232 years), followed by natural gas reserves (64.8 years) and oil (43.1 years). The trend is similar in Asia and Australia. With the inclusion of the Islamic Republic of Iran the oil and natural gas reserve situation improves substantially. Compared with previous estimates made as of the end of 1988, world oil reserves at the end of 1992 show an increase of about 10 per cent. Estimates of reserves for natural gas and coal have also been revised slightly upwards.

In the ESCAP region, nuclear energy is used to generate electricity only in China, India, Japan, Pakistan, the Republic of Korea, and Taiwan Province of China.

The ESCAP region has the highest hydropower potential of all the regions in the world. Table 2 shows the technical hydropower potential of developing countries in various regions of the world. Table 3 shows the current status of hydropower development and potential in countries and areas of the ESCAP region.

[*] Notes by the ESCAP secretariat for the first session of the Committee on Environment and Sustainable Development, Bangkok, 4-8 October 1993 (E/ESCAP/ESD/2 and 4).

[1] The Committee stressed that energy was not only a necessary input to socio-economic development but also crucial and indispensable to that process.

Table 1. World's proven reserves of fossil fuels, as of end 1992

	Oil		Natural gas		Coal	
	Amount (Billions of tons)	R/P ratio (years)	Amount (TCF)	R/P ratio (years)	Amount (Billions of tons)	R/P ratio (years)
World	136.5	43.1	4 885.4	64.8	1 039.2	232
Middle East	89.5	99.6	1 520.1	> 100		
OECD	7.4	9.6	477.2	16.0	438.0	250
OPEC	104.9	81.8	1 965.7	> 100	n.a.	n.a.
Asia and Australasia (including China but excluding the Middle East)	5.9	17.9	341.0	52.5	303.9	179
China	3.2	22.2	49.4	92.6	114.5	n.a.
Iran (Islamic Republic of)	12.7	73.6	699.2	> 100	n.a.	n.a.

Source: *BP Statistical Review of World Energy* (June 1993).

Notes: *Proven reserves* are generally taken to be those quantities which geological and engineering information indicate with reasonable certainty can be recovered in the future from known reservoirs under existing economic and operating conditions.

Reserves/production (R/P) ratio: If the reserves remaining at the end of any year are divided by the production in that year, the result is the length of time that those remaining reserves would last if production were to continue at the then current level.

OECD: Organisation for Economic Cooperation and Development.
OPEC: Organization of the Petroleum Exporting Countries.

Table 2. Hydropower development status in developing countries of the world, by region

Region (1)	Technical capacity (gigawatts) (2)	Potential energy (terawatt hours/year) (3)	Expected installed capacity 1995 (gigawatts) (4)	Expected demand, 1995 (terawatt hours) (5)	Expected exploitation ratio, 1995 (6) = (4)/(2)	Expected ratio of potential to 1995 demand (7) = (3)/(5)
World	1 676	7 802	339	3 004	0.2	2.6
West Africa	85	372	10	85	0.11	4.4
East Africa	216	993	18	63	0.08	15.8
Europe, Middle East and North Africa	100	357	45	540	0.45	0.7
Latin America and the Caribbean	550	2 665	143	916	0.26	2.9
ESCAP region	725	3 415	123	1 400	0.17	2.4
South Asia	171	723	44	381	0.26	1.9
East Asia and the Pacific	554	2 692	79	1 019	0.14	2.6

Source: Based on "A survey of the future role of hydroelectric power in 100 developing countries", World Bank Energy Department Paper No. 17 (August 1984).

Table 3. Water power potential and development in countries and areas of the ESCAP region, 1990

Country or area	Estimated potential of hydropower stations capacity (MW)			Present installed capacity (MW)			Ratio of exploited capacity of estimated potential (percentage)
	Total	Medium and large	Small/mini and micro	Total	Medium and large	Small/mini and micro	
(1)	(2)	(3)	(4)	(5)	(6)	(7)	(8)
Afghanistan	28 000.00	23 000.00 (Amu River at boundary) +5 000.00 (internal rivers)	..	260.85	260.85	..	0.93
Bangladesh	330.00	330.00	..	230.00	230.00	..	69.70
Bhutan	20 050.00	12 050.00	8 000.00[a]		336.00	5.15	1.71
China	680 000.00	660 000.00	20 000.00	36 045.50	22 376.60	13 668.90	4.80
Fiji	322.89	289.00	33.89	83.30	83.20	0.10	26.00
Guam	..	..	0.50	..	..	..	..
India	84 044.00	84 044.00	..	18 307.12	18 084.10	223.02	21.52
Indonesia	75 143.50	73 636.20	1 507.30	2 095.20	1 616.80	478.40	..
Islamic Republic of Iran	16 000.00	15 880.96	119.04	1 952.50	1 936.50	16.00	12.20
Japan	48 042.00	..	..	26 475.00	..	..	55.10
Lao People's Democratic Rep.	18 000.00	..	..	150.00	..	..	0.83
Malaysia Peninsular	4 000.00	3 057.00	943.00	1 251.00	1 230.00	21.00	31.30
Malaysia Sabah	..	..	..	..	66.00	4.50	..
Malaysia Sarawak	107.30	..	107.30	109.73	107.33	2.40	0.50
Myanmar	24 736.70	24 623.20	113.50	257.86	249.00	8.86	1.04
Nepal	..	..	..	232.69	221.20	11.49	..
New Zealand	8 582.00	8 582.00	..	4 639.00	4 534.00	105.00	54.10
Pakistan	30 700.00	30 000.00	700.00	2 897.00	2 790.00	0.00	9.40
Papua New Guinea	25 000.00	..	..	230.20	202.20	28.00	0.90
Philippines	11 677.00	10 861.00	816.00	2 204.00	2 198.20	5.80	18.90
Republic of Korea	3 500.00	3 177.00	323.00	1 340.00	1 311.00	29.00	38.30
Solomon Islands	8.00	..	..	--	--	--	0.00
Territory of American Samoa	0.06	..	..	--	--	--	0.00
Sri Lanka	870.00	870.00	..	1 115.00	1 115.00	..	..
Thailand	..	..	..	2 273.70	2 239.70	33.40	..
Vanuatu	3 500.00	--	3 500.00	--	--	--	0.00

Source: Based on the forthcoming United Nations publication, *Electric Power in Asia and the Pacific, 1989 and 1990* (ST/ESCAP/1286).

Note: [a] Unidentified.

Two dots (..) means not available.

Two dashes (..) means negligible.

II. ENERGY SUPPLIES

A. Commercial primary energy

Commercial primary energy is produced from solids, liquids, gas and electricity. Solids comprise hard coal, lignite, peat and oil shale; liquids comprise crude petroleum and natural gas liquids; gas comprises natural gas; and electricity comprises primary electricity generated from hydropower, nuclear and geothermal sources.

Table 4 shows the production of commercial primary energy in the ESCAP region excluding the newly independent Asian republics. The production pattern of commercial energy was erratic in 1973-1980, the period during which two major oil price shocks occurred. At the end of the period the production level was only slightly higher than in 1973, indicating an average annual growth of only 0.25 per cent. This low growth rate was exclusively due to the decline of liquid fuels production from 1979 to 1981. The growth, however, picked up quite strongly thereafter, achieving a steady average annual rate of 5.86 per cent over the period 1980-1990. At the end of the period the total primary energy production in the region reached over 1,515 million tons of oil equivalent (TOE).

Table 4. Production of commercial energy in the ESCAP region[a]

(Million tons of oil equivalent)

	1973	1980	1985	1986	1987	1988	1989	1990	Average annual growth rates (percentage)		
									1980/1973	1990/1980	1990/1973
Solids	319.5 (37.9)	437.8 (51.1)	631.9 (53.9)	660.2 (54.4)	726 (55.6)	752.7 (55.5)	802.7 (55.2)	824.2 (54.4)	4.6	6.53	5.73
Liquids	466.0 (55.3)	322.4 (37.6)	395.7 (33.7)	396.1 (32.6)	411.0 (31.5)	420.9 (31.0)	453.8 (31.2)	479.6 (31.7)	−5.13	4.05	0.17
Gas	39.9 (4.7)	66.9 (7.8)	101.5 (8.7)	112.1 (9.2)	118.3 (9.1)	128.7 (9.5)	140.8 (9.7)	152.6 (10.1)	7.66	8.6	8.21
Electricity	17.0 (2.0)	30.0 (3.5)	43.5 (3.7)	45.9 (3.8)	50.7 (3.9)	53.6 (4.0)	56.0 (3.9)	58.8 (3.9)	8.45	6.96	7.57
Total	842.4 (100)	857.1 (100)	1 172.6 (100)	1 214.3 (100)	1 306.0 (100)	1 355.9 (100)	1 453.3 (100)	1 515.2 (100)	0.25	5.86	3.51

Source: *United Nations Yearbook of World Energy Statistics,* various issues; and *1990 Energy Statistics Yearbook* (United Nations publications, Sales No. E./F.92.XVII.3).

Note: The figures in the parentheses show the share as a percentage of the total.

[a]   Excluding the newly independent Asian republics.

B. Non-commercial energy

Certain types of traditional sources of energy, such as fuelwood, are considered non-commercial energy although some trading in them does occur. Renewable sources in general, except hydropower and geothermal, fall under this category.

Non-commercial energy plays an important role in the energy supply, particularly in rural areas, in developing countries of the ESCAP region. In some countries the share of non-commercial energy in the energy supply is over 80 per cent. However, for lack of comparable and reliable data, it is not possible to document exact amount of such energy production in the region.

A TCDC programme to promote the accelerated use of new and renewable sources of energy to augment the rural energy supply is being implemented by the secretariat and several countries of the region with and supplementary funding support from donor countries. The programme needs additional support from donor countries and agencies and active participation from developing countries.

III. HISTORICAL ENERGY DEMAND AND FUTURE PERSPECTIVE

A. Regional aggregated commercial energy consumption trends

Commercial energy is consumed in the form of solids, liquids, gases and electricity. The consumption of solids refers to the use of primary forms of solid fuels, net imports and changes in stocks of secondary fuels; the consumption of liquids refers to the use of petroleum products (including feedstocks), natural gasoline, condensate and refinery gas and to inputs of crude petroleum to thermal power plants; the consumption of gases refers to the use of natural gas and coke-oven gas, net imports and changes in stocks at gas works; and the consumption of electricity refers to the use of primary electricity and net imports of electricity.

Table 5 shows the historical trend of consumption of all forms of commercial primary energy in the ESCAP region and aggregate world trends. Regional primary energy consumption increased at an average annual rate of 4.45 per cent over the period 1973-1990, a significant growth rate compared with the world average growth of 1.92 per cent. The regional total reached 1,637.1 million TOE, up from 781.5 million TOE in 1973. The increase can be attributed almost entirely to the high growth of energy consumption in developing economies of the region. Except for a slight drop in consumption in 1980, the consumption in developing economies has been rising continuously. During the 1980s the average annual growth rate was about 6.2 per cent in developing economies of the region, compared with less than 2.2 per cent in developed economies of the region. The higher growth in developing economies has been attributed to the low energy consumption level but

Table 5. Consumption of commercial primary energy
(Million tons of oil equivalent and kilograms of oil equivalent per capita))

	1973	1980	1985	1986	1987	1988	1989	1990	Average annual growth rates (percentage)		
									1980/1973	1990/1980	1990/1973
World	5 206.7 (1 346)	5 996.3 (1 362)	6 436.4 (1 327)	6 520.4 (1 321)	6 773.9 (1 345)	7 034.9 (1 373)	7 190.1 (1 378)	7 199.8 (1 352)	2.04	1.85	1.92
ESCAP region	781.5 (369)	1 017.0 (410)	1 256.5 (466)	1 287.5 (468)	1 385.7 (495)	1 483.2 (520)	1 577.7 (544)	1 637.1 (551)	3.83	4.88	4.45
ESCAP developed countries	340.4 (2 722)	370.1 (2 748)	403.0 (2 867)	395.4 (2 810)	399.8 (2 825)	426.7 (3 000)	439.8 (3 076)	459.3 (3 079)	1.2	2.18	1.78
ESCAP developing countries	441.0 (220)	646.9 (279)	853.5 (334)	892.1 (342)	985.9 (371)	1 056.5 (390)	1 137.9 (412)	1 177.9 (417)	5.62	6.18	5.95

Source: *United Nations Yearbook of World Energy Statistics,* various issues; and *1990 Energy Statistics Yearbook* (United Nations publications, Sales No. E./F.92.XVII.3).

Note: The figures in the parentheses show per capita consumption.

higher than world average economic growth rate, higher than average population growth, rapid urbanization, high energy intensity due to economic development, and inefficient energy use.

The region has been a net importer of energy since 1979; in 1990, it produced 1,515 million TOE but consumed 1,637 million TOE.

Although the per capita energy consumption in the developing economies of the ESCAP region has been increasing, it remains low. In 1990 the average per capita commercial energy consumption in the developing economies of the ESCAP region reached only 417 kilograms (kg) of oil equivalent. The regional average was 551 kg and the world average 1,352 kg of oil equivalent. The situation in individual countries varies. Many developing economies depend on non-commercial forms of energy to meet bulk of their energy demand. Nevertheless, the overall energy consumption in developing economies of the ESCAP region is low.

The consumption pattern in terms of energy mixes of commercial primary energy in the ESCAP region is shown in table 6. Solid fuels have had the largest share of consumption in the region since 1980 and over a 50 per cent share in total commercial energy consumption since 1984. The share of liquid fuels decreased significantly, from a high of 50 per cent in 1973 to 33.9 per cent in 1987, though it increased modestly to 34.7 per cent in 1990. The decrease in the share of liquid fuels since 1978 contrasts with the increase in the share of solid fuels and gaseous fuels. The share of primary electricity increased slightly, from

2.2 per cent in 1973 to 3.5 per cent in 1985, remaining at approximately the same level since then. Tables 7 and 8 show the commercial energy consumption patterns in developed and developing economies. China, as both the largest producer and consumer of energy among the developing economies, influences the consumption pattern of the developing economies in the region. Table 9 shows the breakdown for 1990 of the consumption pattern of commercial primary energy in the developing economies of the ESCAP region. Although solids appear to be the dominant fuel (62.7 per cent) for the region as a whole, when China is excluded the share of solids drops to 38.9 per cent and the share of liquids jumps to 44.1 per cent.

B. Sectoral energy demand trends in selected economies of the ESCAP region

Table 10 shows the trend of sectoral energy demand in selected developing countries of the ESCAP region for the industry, transport, and household and others sectors. The most significant trend observed is the increase of the share of the transport sector in energy demand in most of the countries. Upward changes are observed in some countries in the "household and others" sector, which may be because of either the influence of "other" economic activities (such as commerce, services) in that sector or the significant improvement in the standard of living. The industry sector in developing economies shows a mixed trend (some ups and some downs) in energy consumption as a percentage of total consumption, which signifies structural adjustments in economic activities.

6

Table 6. Consumption pattern of commercial primary energy in the ESCAP region

(Million tons of oil equivalent)

	1973	1980	1985	1986	1987	1988	1989	1990
Solids	342.3	465.4	655.0	680.1	743.9	788.3	839.5	857.9
	(43.8)	(45.76)	(52.13)	(52.66)	(53.68)	(53.15)	(53.21)	(52.4)
Liquids	390.9	453.9	453.5	452.6	469.5	508.0	535.9	567.5
	(50.02)	(44.63)	(36.09)	(35.04)	(33.88)	(34.25)	(33.97)	(34.66)
Gas	31.3	67.8	104.5	112.8	121.7	133.6	146.6	153.2
	(4.01)	(6.67)	(8.32)	(8.73)	(8.78)	(9.01)	(9.29)	(9.36)
Electricity	17.0	30.0	43.5	46.2	50.7	53.6	55.9	58.8
	(2.18)	(2.95)	(3.46)	(3.58)	(3.66)	(3.61)	(3.54)	(3.59)
Total	781.5	1 017.0	1 256.5	1 291.6	1 385.7	1 483.2	1 577.7	1 637.1
	(100)	(100)	(100)	(100)	(100)	(100)	(100)	(100)

Source: *United Nations Yearbook of World Energy Statistics,* various issues; and *1990 Energy Statistics Yearbook* (United Nations publications, Sales No. E./F.92.XVII.3).

Note: The figures in the parentheses show the share as a percentage of the total.

Table 7. Consumption pattern of commercial primary energy in developed countries of the ESCAP region

(Million tons of oil equivalent)

	1973	1980	1985	1986	1987	1988	1989	1990
Solids	75.6	84.6	109.1	104.8	106.5	114.6	117.5	119.3
	(22.21)	(22.86)	(27.07)	(26.49)	(26.64)	(26.86)	(26.72)	(25.97)
Liquids	247.0	235.4	216.2	208.7	209.1	224.8	230.7	242.6
	(72.56)	(63.6)	(53.65)	(52.76)	(52.3)	(52.68)	(52.46)	(52.82)
Gas	8.8	32.3	53.2	56.8	58.0	60.4	64.0	68.3
	(2.5)	(8.73)	(13.2)	(14.36)	(14.51)	(14.16)	(14.55)	(14.87)
Electricity	9.4	17.8	24.5	25.3	26.3	26.9	27.7	29.1
	(2.76)	(4.81)	(6.08)	(6.4)	(6.58)	(6.3)	(6.3)	(6.34)
Total	340.4	370.1	403.0	395.6	399.8	426.7	439.8	459.3
	(100)	(100)	(100)	(100)	(100)	(100)	(100)	(100)

Source: *United Nations Yearbook of World Energy Statistics,* various issues; and *1990 Energy Statistics Yearbook* (United Nations publications, Sales No. E./F.92.XVII.3).

Note: The figures in the parentheses show the share as a percentage of the total.

Table 8. Consumption structure of commercial primary energy in developing countries of the ESCAP region

(Million tons of oil equivalent)

Fuel type/years	1973	1980	1985	1986	1987	1988	1989	1990
Solids	266.8	380.8	545.9	575.2	637.4	673.6	722.0	738.6
Percentage	(60.5)	(58.9)	(64.0)	(64.2)	(64.7)	(63.8)	(63.5)	(62.7)
Liquids	143.9	218.5	237.3	243.9	260.4	283.2	305.2	324.9
Percentage	(32.6)	(33.8)	(27.8)	(27.2)	(26.4)	(26.8)	(26.8)	(27.6)
Gas	22.8	35.5	51.3	55.9	63.8	73.2	82.6	84.9
Percentage	(5.2)	(5.5)	(6.0)	(6.2)	(6.5)	(6.9)	(7.3)	(7.2)
Electricity	7.6	12.2	19.0	20.9	24.4	26.7	28.3	29.7
Percentage	(1.7)	(1.9)	(2.2)	(2.3)	(2.5)	(2.5)	(2.5)	(2.5)
Total	441.0	646.9	853.5	895.9	985.9	1 056.6	1 137.9	1 177.9
Percentage	(100)	(100)	(100)	(100)	(100)	(100)	(100)	(100)

Source: United Nations Yearbook of World Energy Statistics, various issues; and *1990 Energy Statistics Yearbook* (United Nations publications, Sales No. E./F.92.XVII.3).

Note: The figures in the parentheses show the share as a percentage of the total; the total may not add up owing to rounding and conversion of units.

Table 9. Consumption pattern of commercial primary energy in developing economies of the ESCAP region, breakdown for 1990

(Millions of tons of oil equivalent)

	ESCAP region	Excluding China	Excluding China and India
Solids	738.6	207.1	82.8
	(62.7)	(38.9)	(23.9)
Liquids	324.9	234.7	189.1
	(27.6)	(44.1)	(54.5)
Gas	84.9	70.7	61.4
	(7.2)	(13.3)	(17.7)
Electricity	29.7	20.0	13.8
	(2.5)	(3.8)	(4.0)
Total	1 177.9	532.5	347.1
	(100)	(100)	(100)

Source: United Nations Yearbook of World Energy Statistics, various issues; and *1990 Energy Statistics Yearbook* (United Nations publications, Sales No. E./F.92.XVII.3).

Note: The figures in the parentheses show the share as a percentage of the total.

Table 10. Sectoral energy demand[a] pattern in selected developing countries of the ESCAP region

Country or area	Year	Sectors			All sectors
		Industry	Transport	Household and others	
Bangladesh	1980	31.70	18.91	49.39	100
	1990	26.06	14.04	59.90	100
Fiji	1980	23.22	56.87	19.91	100
	1990	13.48	69.50	17.02	100
India	1980	52.46	24.82	22.72	100
	1990	52.47	22.33	25.20	100
Indonesia	1980	32.23	29.02	38.75	100
	1990	30.27	29.36	40.37	100
Malaysia	1980	44.95	37.56	17.49	100
	1990	40.13	40.97	18.90	100
Philippines	1980	52.74	27.15	20.11	100
	1990	39.67	37.65	22.68	100
Republic of Korea	1980	38.70	13.98	47.32	100
	1990	39.94	19.74	40.32	100
Singapore[b]	1980	27.22	56.22	16.56	100
	1990	37.01	47.96	15.03	100
Sri Lanka	1980	22.87	48.52	28.61	100
	1990	18.57	54.96	26.47	100
Thailand	1980	28.89	41.96	29.14	100
	1990	25.20	52.15	22.66	100

Sources: Asian Development Bank, *Energy Indicators of Developing Member Countries of Asian Development Bank* (Manila, July 1992).

International Energy Agency, *Energy Statistics and Balances of Non-OECD Countries,* Paris, 1979-1987, 1988-1989.

Notes: [a] Final energy consumption, including non-commercial energy.

[b] IEA Energy Statistics and Balances of Non-OECD Countries.

C. Future perspective

It is widely expected that the historical trend observed during the 1980s will continue in the future and that the Asian and Pacific region will diversify its energy mix and thereby become less and less dependent on oil. It is also expected that, despite its poor image as a polluting fuel, coal will be increasingly consumed. Technological advances will mitigate some of the harmful environmental impact of coal utilization.

The forecasts analyzed below are intended to provide a plausible global framework of the development of energy trends. Several studies have been made pertaining to future developments with regard to regional energy. For instance, the Intergovernmental Panel on Climate Change had commissioned working groups to develop projections of future energy demand to determine the contribution of energy to CO_2 emissions. The secretariat used two projections, one from the "Energy exploration and development trends in

developing countries: report of the Secretary-General,"[1] and one from "Energy Policy Implications of Climatic Effects of Fossil Fuel Use in the Asia-Pacific Region",[2] which produced projections of energy demand to the year 2010.

The following analysis is based on projections from the two sources.

1. Report of the Secretary-General

Table 11 shows for various regions of the world the level of commercial primary energy consumption for different years as well as the estimated breakdown of the demand in terms of four principal energy types (oil, gas, coal and electricity). By the year 2010 the energy consumption in developing countries will be

---

[1] A/45/274-E/1990/73.

[2] ST/ESCAP/1007.

Table 11. Estimated future consumption of commercial primary energy[a]

(Million tons of oil equivalent)

		1988	1995	2000	2010	Annual growth (percentage)
Developed market economies	Oil	1 596	1 653	1 775	2 048	1.2
	Gas	749	857	923	1 070	1.6
	Coal	915	995	1 086	1 294	1.8
	Electricity	228	354	451	686	5.2
	Total	3 488	3 859	4 235	5 098	1.8
Eastern Europe and the USSR[b]	Oil	466	558	654	603	3.0
	Gas	617	833	941	1 188	2.9
	Coal	664	705	765	899	1.6
	Electricity	45	58	114	273	8.5
	Total	1 793	2 154	2 474	2 963	2.8
All developing countries	Oil	683	899	1 124	1 759	4.4
	Gas	245	342	427	678	4.8
	Coal	730	966	1 163	1 686	3.8
	Electricity	70	140	197	353	7.5
	Total	1 728	2 347	2 910	4 476	4.4
Member countries of OPEC	Oil	152	193	237	357	4.0
	Gas	117	155	177	231	3.1
	Coal	4	6	7	10	4.3
	Electricity	4	17	20	18	7.0
	Total	278	370	441	615	3.7
Other oil exporting countries	Oil	216	291	356	536	4.2
	Gas	65	92	116	186	4.9
	Coal	477	629	751	1 072	3.8
	Electricity	17	33	50	103	8.5
	Total	774	1 044	1 274	1 897	4.2
Oil-importing countries	Oil	315	416	531	867	4.7
	Gas	63	96	134	261	6.7
	Coal	250	331	404	604	4.1
	Electricity	49	90	126	233	7.3
	Total	676	933	1 196	1 964	5.0
Total world	Oil	2 746	3 110	3 553	4 410	2.5
	Gas	1 611	2 032	2 291	2 936	2.7
	Coal	2 309	2 666	3 014	3 879	2.5
	Electricity	343	552	762	1 312	6.3
	Total	7 009	8 360	2 619	12 537	2.8

Source: *Energy exploration and development trends in developing countries, Report of the Secretary General* (A/45/274-E/1990/73).

Notes: [a] In the baseline scenario, crude oil prices (in constant 1989 United States dollars) are assumed to remain constant at $18 per barrel throughout the projection period. The annual rate of economic growth is estimated to average 3.1 per cent for the developed market economies, 3.1 per cent for eastern Europe and the Soviet Union, 4.2 per cent for OPEC member countries, 3.9 per cent for other oil-exporting developing countries and 4.7 per cent for the oil-importing developing countries.

[b] Predictions made for the former Union of Soviet Socialist Republic (USSR).

OPEC – Organization of the Petroleum Exporting Countries.

approaching that in the developed market economies. The developing countries by 1995 or sooner will have surpassed the energy consumption of eastern Europe and the former Union of Soviet Socialist Republics. According to the projection, by the year 2010 world energy consumption will rise to over 12,500 MTOE (million tons of oil equivalent) representing an annual average growth rate of 2.8 per cent. The comparable figures for developing countries are 4,476 MTOE and 4.4 per cent.

2. ESCAP climatic effects study

A study was undertaken by ESCAP in 1989-1990, with the generous support of the Government of Japan, on the energy policy implications of increased fossil fuel utilization. The results of the study were discussed in a seminar held in Tokyo in December 1990. The following three scenarios were developed:

(a) *Business as usual.* Economic trends in 1973-1986 continue into the future with gradually diminishing growth rates. Awareness of the need for energy conservation measures grows, but no substantial investments are made in efficient energy use equipment.

(b) *Energy efficiency.* Economic growth rates are the same as in the above scenario, but more concerted efforts by government are responsible for achieving a 20-25 per cent reduction in energy consumption. Thus investments in efficient energy use equipment are made on their economic merit alone.

(c) *Conservation with fuel-switching.* With economic growth rates the same as before and an efficient energy scenario, additional investments are made to switch fuels from coal to gas or at least to oil, and from fossil fuel to non-fossil fuel, with a net additional economic cost of abatement of the impact on the environment.

In the business-as-usual scenario, the annual growth rate of commercial energy consumption is projected to be 4.7 per cent for the period 1986-2000 and 4.4 per cent for the period 2000-2010. In the energy efficiency scenario, the annual growth rate is 3.7 per cent for the period 1986-2000 and 3.4 per cent for the period 2000-2010; similar growth is projected for the conservation with fuel-switching scenario.

When compared with the historical annual growth rate of 6.2 per cent in commercial energy consumption by developing economies of the ESCAP region during the 1980s, the projected future demand for commercial

energy is considered somewhat conservative. However, lower economic growth rates and expected energy conservation efforts in some developing countries may justify lower energy demand growth over the next two decades. Although some adjustments may have since been made in the projections for some countries as a result of changes in the global socio-economic situation, the projected ESCAP regional energy demand growth of 4.7 and 4.4 in the base case scenario in 1986-2000 and 2000-2010 respectively compares well with the overall developing country growth scenario of 4.4 per cent for the period 1988-2010 as shown in table 11.

The projections in figures I-IV, take into account the different shares of the principal energy types. Most energy consumed in the Asian and Pacific region has been and will continue to be from solid fuels because the regional trend is dominated by two large economies, China and India, with substantial coal resources. If both of these economies are taken out of the projections, then most of the energy demand will be for liquid fuels. Energy mixes are of course different for the different subregions and economies.

The relation between gross domestic product (GDP) and commercial energy consumption may be used to check the overall energy projection. This relation is said to have "decoupled" since the oil crisis of 1973-1974. There are indications, however, that a certain "recoupling" has occurred since the oil price crash of 1986. This confirms the view that the coupling holds as long as there are no price increases (or if the energy price slowly decreases). On the basis of this relationship it was determined that the results of the climatic effects study were on the whole lower than the figures thus obtained, perhaps because of an implicit assumption of a slowly increasing energy price in the climatic effects study. Alternatively, it may be that the assumption of a constant energy price over a long period (in this case 1986-2010) results in very high energy demand projections that are likely to be unsustainable. Hence the inescapable conclusion is that a constant energy price scenario is unsustainable.

Similar high growth concerns apply to the electricity demand. In fact, the current electricity supply situation in some countries (Indonesia (Java), the Philippines and Thailand) is vulnerable to interruptions because of the lack of reserve capacity. Large investments are urgently required: Thailand, for example, allocated 80 billion baht (more than US$ 3 billion) to the energy (public) sector in 1992. A similar level of expenditure is required in Indonesia and the Philippines to overcome shortages.

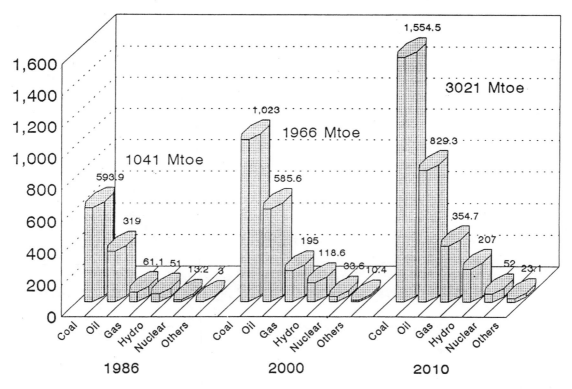

Source: *Energy Policy Implications of the Climatic Effects of Fossil Fuel Use in the Asia-Pacific Region*

Figure I. Projection of total energy consumption:
Scenario S1 (business as usual) Asia-Pacific region

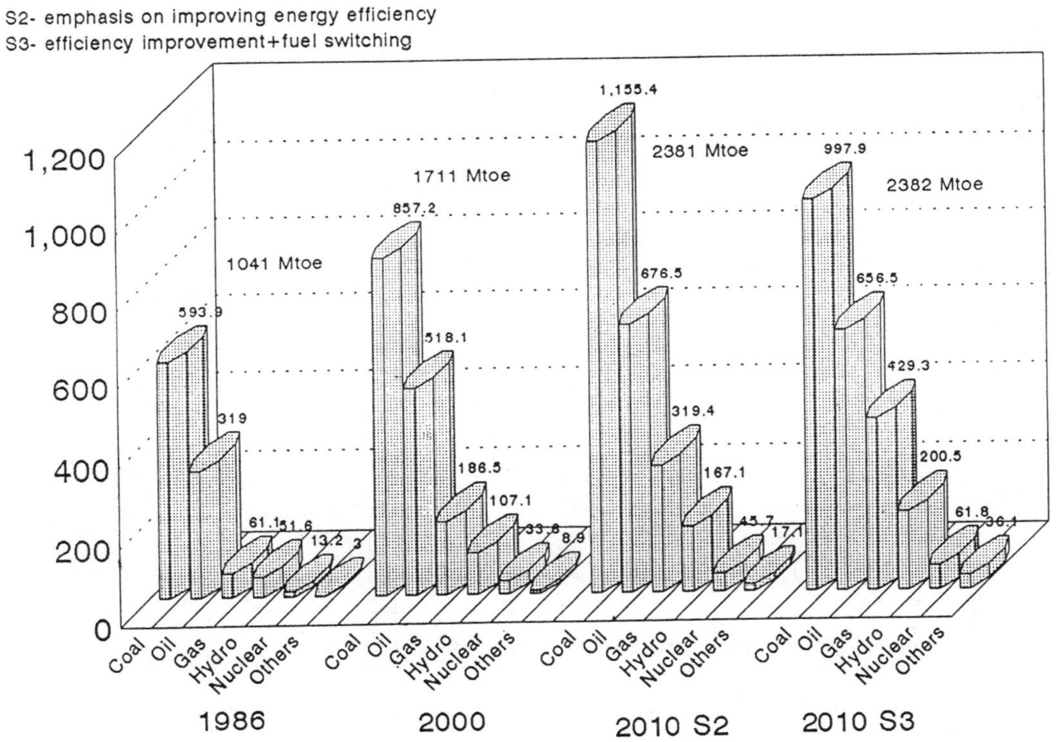

S2- emphasis on improving energy efficiency
S3- efficiency improvement+fuel switching

Source: *Energy Policy Implications of the Climatic Effects of Fossil Fuel Use in the Asia-Pacific Region, ESCAP, 1991.*

Figure II. Projection of total primary energy consumption:
Scenario S2 and S3 Asia-Pacific region

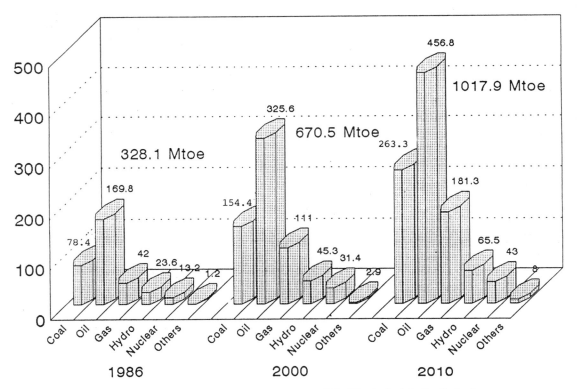

Source: Energy Policy Implications of the Climatic Effects of Fossil Fuel Use in the Asia-Pacific Region

Figure III. Projection of total primary energy consumption:
Scenario S1 (business as usual) Asia-Pacific region excluding India and China

S2-emphasis on improving energy efficiency
S3-efficiency improvement+fuel Switching

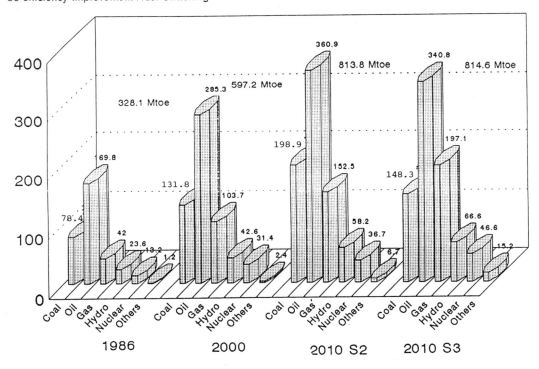

Source: Energy Policy Implications of the Climatic Effects of Fossil Fuel Use in the Asia-Pacific Region, ESCAP, 1991.

Figure IV. Projection of total primary energy consumption:
Scenario S2 and S3 Asia-Pacific region excluding India and China

IV. ENERGY ISSUES AND POLICIES

A. Supply and demand

1. Supply

Choosing an appropriate energy mix for an economy is a difficult task. Apart from techno-economic considerations, environmental policy is becoming a key factor in determining the energy mix. "Indigenization" of energy supply is another policy option that influences the choice of energy mix, as do policies on diversity and security of supply.

Environmental concern have complicated the choice of energy sources. The decision to develop resources like hydropower often sets off environmental protests even though hydropower is a renewable resource. Similarly, there is increasing concern about pollution from burning fossil fuels, particularly coal. Nuclear power development also faces considerable resistance in many countries that have the potential for such development.

Natural gas, an environmentally benign resource, is being used in well-endowed countries. Unfortunately, owing to high infrastructural costs, transport and handling problems, the international trade in natural gas remains limited. Trade in liquified natural gas (LNG), and through pipelines has been limited by the huge infrastructural and investment requirements of such trade. Regional/subregional cooperation in integrating demand and supply interests could be helpful. Interest exists in projects for building intercountry pipelines in South-East Asia, South-West Asia and the central Asian republics subregions, as well in East Asia.

Augmentation and improvement of energy supply to rural areas is an important issue in the developing countries of the ESCAP region. An appropriate rural energy mix of commercial and traditional sources of energy is critical for a sustainable energy supply in rural areas that avoids deforestation or desertification and preserves well-being.

2. Demand

Traditionally energy demand has been met by supply side adjustment. Owing to rapid energy demand growth coupled with high energy prices, the concept of demand-side management is slowly being adopted in a number of countries. This is important because the amount of energy used in many countries to produce a given unit of economic output (measured in barrels or tons of oil equivalent used per US$ 1,000 of real GDP) or the intensity of energy use is relatively higher than what is ideally needed. Energy intensity is a measure of the level of economic development, the structure of the economy, existing technologies, and the energy mix.

Although energy demand is driven mainly by real GDP, other important factors include: (a) changes in the structure of the economy because different sectors have different levels and patterns of energy consumption; (b) degree of urbanization as it affects household demand; and (c) availability of different sources of primary energy and forms of secondary energy. These factors also indicate the level of development of an economy.

According to the analyses presented in section III above, energy demand in the ESCAP region, particularly in the developing economies, is expected to grow faster than the growth of GDP. Therefore, conscious effort is necessary to undertake energy demand management measures.

The measures considered for demand management may be classified into the following categories: (a) conservation; (b) interfuel substitution; and (c) interfactor substitution. A strategy may comprise a mix of these measures depending on the situation in the individual country situation and its energy policy directives.

Both pricing and non-pricing policies may be used as implementation tools in demand management. The response of consumers to changes in energy prices can be measured by estimating the price elasticity of demand and elasticity coefficients for interfuel and interfactor substitution, on the basis of historical data.

However because reliable segregated sectoral energy demand data are unavailable in developing countries, the development of a database is a prerequisite to any meaningful analysis of the energy demand and should be followed by an analysis of the possible future evolution of interactions between the sectors.

B. Pricing and financing issues

Financing is the most pressing issue for the exploration, development and management of energy resources, particularly in oil-importing developing countries. The major share of the energy sector investment goes to power sector.

According to a survey[3] of the expansion programmes planned by electric utilities of developing countries, capital expenditure for power generation in these countries during the 1990s has been estimated at US$ 745 billion (at 1989 dollar prices). With price escalation, the total financing requirements may be of the order of US$ 1 trillion. The requirements of the developing countries of the region are estimated at US$ 635 billion.

The capital expenditure has been estimated on the basis of the expansion programmes of individual countries. The average capital expenditure for the additional system capacity in the 1990s has been estimated to be US$ 1,942 per kilowatt (kW) of system capacity added, including transmission and distribution facilities.

The foreign currency component of the financing needs amounts to US$ 222 billion for the developing countries of the ESCAP region. This huge amount of capital financing will have to come from limited sources, including international and regional funding agencies such as the World Bank and the Asian Development Bank (ADB) and bilateral assistance. The average lending to the energy sector by the World Bank (including the International Development Agency (IDA) assistance) and ADB during the 1980s covered only around 10 per cent of the amount needed. The gap between the demand and supply of capital is becoming wider, creating increasing pressure on the financial markets.

The above estimate does not include pollution abatement costs. A World Bank indicative estimate[4] shows that for developing countries additional costs could be in the range of US$ 7 billion in 2000 for coal-fired power stations alone. Table 12 shows the estimated costs and long-term benefits of selected environmental programmes in developing countries.

One of the recent trends observed in the power supply industry of developing countries is the tendency to allow private sector participation in power development activities. One main reason for this is to be able to mobilize the additional financial resources. Private sector investors have their own set of objectives, including the financial objective of an adequate rate of return on their investment. As with investment in any other business, private sector investors will assess the

risk of the investment and compare it with the opportunity costs of their capital. If the investment is riskier than other alternatives, a higher return including a risk premium would be expected. Thus it appears that only utilities with a good operating record and financial health will be able to attract private sector participation. Moreover, the private sector is expected to become involved gradually and will not be able to meet the huge financing demand. Nevertheless, private sector participation can become a significant supplementary source of funding. Participation of the private sector can be expected in such projects as (a) cogeneration; (b) captive generation; (c) build-operate-transfer/build-own-operate; and (d) privatization of existing utilities. Investment in environment related technologies may facilitate both technology transfer and capital flow.

C. Energy conservation and efficiency

Notwithstanding the current energy supply situation, the rational use of energy is important for long-term economic and environmental benefit. From an economic standpoint, increasing the efficiency of energy use is normally more attractive than investing additional resources to increase the energy supply. In addition, energy conservation through reduced demand can prolong the supply and make it available instantaneously. Experience amply demonstrates that cost-effective conservation measures can save the energy and foreign exchange of developing countries by reducing energy imports. Moreover, energy conservation can lead to the mitigation of the problems of environmental degradation and global warming.

National energy conservation policies should be developed and implemented to encourage energy conservation. While some awareness of the need for energy conservation has been created in developing countries of the ESCAP region, implementation of conservation measures has so far been very slow, partly because of the lack of definite policies and regulatory measures. Even in countries in which a legal framework has been established, its effectiveness is seldom measured because of financial, administrative and manpower constraints. A separate article "Prospects for Energy Sector Implementation of Agenda 21: The Role of Energy Efficiency" discusses these issues in greater details and includes a proposal for an Asian energy efficiency programme for the twenty-first century.

D. Environmental issues

Energy systems, as well as their production, conservation and utilization, have some adverse impacts on the environment. There is a growing awareness of the need to mitigate these impacts as much as possible.

---

[3] *Capital Expenditures for Electric Power in the Developing Countries in the 1990s,* Industry and Energy Department Working Paper, Energy Series Paper No. 21 (World Bank, February 1990).

[4] See table 12 of the present document.

Table 12. Estimated costs and long-term benefits of selected environmental programmes in developing countries

Programme	Additional investment in 2000			Long-term benefits
	Billions of dollars a year	*As a percentage of GDP in 2000[a]*	*As a percentage of GDP growth, 1990-2000[a]*	
Increased investment in water and sanitation	10.0	0.2	0.5	Over 2 billion more peple provided with service. Major labor savings and health and productivity benefits. Child mortality reduced by more than 3 million a year.
Controlling particulate matter (PM) emissions from coal-fired power stations	2.0	0.04	0.1	PM emissions virtually eliminated. Large reductions in respiratory illnesses and acid deposition, and improvements in amenity.
Reducing acid deposition from new coal-fired stations[b]	5.0	0.1	0.25	
Changing to unleaded fuels; controls on the main pollutants from vehicles[b]	10.0	0.2	0.5	Elimination of pollution from lead; more than 90 per cent reductions in other pollutants, with improvements in health and amenity.
Reducing emissions, effluents, and wastes from industry	10.0-15.0	0.2-0.3	0.5-0.7	Appreciable reductions in levels of ambient pollution, and improvements in health and amenity, despite rapid industrial growth. Low-waste processes often a source of cost savings for industry.
Soil conservation and afforestation, including extension and training	15.0-20.0	0.3-0.4	0.7-1.0	Improvements in yields and productivity of agriculture and forests, which increase the economic returns to investment. Lower pressures on natural forests. All areas eventually brought under sustainable forms of cultivation and pasture.
Additional resources for agricultural and forestry research, in relation to projected levels, and for resource surveys	5.0	0.1	0.2	
Family planning (incremental costs of an expanded program)[c]	7.0	0.1	0.3	Long-term world population stabilizes at 10 billion instead of 12.5 billion.
Increasing primary and secondary education for girls[c]	2.5	0.05	0.1	Primary education for girls extended to 25 million more girls, and secondary education to 21 million more. Discrimination in education substantially reduced.

Source: World Bank, *World Development Report 1992* (New York, Oxford University Press, 1992).

Notes: [a] The GDP of developing countries in 1990 was $ 3.4 trillion, and it is projected to rise to $ 5.4 trillion by 2000 (in 1990 prices). The projected GDP growth rate is 4.7 per cent a year.

[b] Costs may eventually be lowered by the use of new combustion technologies and other measures.

[c] Recurrent expenditures on these items are counted as investments in human resources.

Energy planners and policy makers can no longer avoid environmental considerations. They will have to find and develop a strategy to satisfy energy demand and to sustain the environment. Suitable policies and instruments are needed to implement such policies.

National policy and regional cooperation requirements for energy and the environment should be defined. Resource constraints in developing countries could be an obstacle in implementing some of the measures.

V. SUSTAINABLE ENERGY DEVELOPMENT

During the past several years the use of the words "development" and "sustainable" has been quite frequent. In the latest global forum, the United Nations Conference on Environment and Development (UNCED), the world leaders accepted "Agenda 21" as the strategy for future development activities in areas of socio-economic development. Energy, is an important input to economic development, and therefore the adequate and ensured supply of energy is a concern in all countries. Although environmental problems have been discussed increasingly since the early 1970s, it is only recently that the sustainability of the environment has become a key issue. This issue, however, is a cause of more immediate concern for the industrialized countries than for the developing countries. The immediate concern for developing countries is how to meet their growing demand for energy. As almost all the commercial energy sources are finite, it is important to find ways and means to keep them from being depleted. One analyst has described sustainability as follows: "If we can find a material "X" that would be available for only 50 years, but which will leave no serious long-term residuals for future generations, it would be a highly desirable energy source. Before the 50 years are up, we might be able to find another material "Y" that could meet the same criteria for acceptability".[5]

To make energy development sustainable several options should be considered either singly or in combination. These options can be broadly grouped into two categories: (a) supply options and (b) energy efficiency improvement.

A. Supply options

In the supply option, a couple of measures can be taken: to ensure future supplies and to opt for or switch to sources of energy that produce less pollution per unit of useful energy. New sources of energy, including renewable energy, increase future supplies, and switching from coal and oil to natural gas and further to hydropower, nuclear and other renewable sources would result in the reduction of emissions such as carbon and sulfur dioxides, nitrogen oxides, and particulate emissions, although other types of environmental problems (from some large hydropower plants) and potential problems (radioactive emissions from nuclear plans) may occur.

Most of the environmental impacts of energy chain are local in nature, affecting water and air. This has serious economic, environmental and health impacts on the local community.

In the areas where energy use is high and sources of emissions are closer to other countries, the trans-boundary impact of pollution (acid rain) becomes a regional issue.

Climatic change or global warming, caused among other factors by the accumulation of greenhouse gases such as CO_2 in the atmosphere is a matter of concern for the whole world. Although the effect of greenhouse gases in terms of climatic changes is not known with certainty, the danger of ignoring the issue could prove catastrophic. The issue of greenhouse gas emissions has become an increasing source of contention between the developing and developed economies, especially with regard to the issues of priority-setting and the sharing of responsibility for the mitigation or prevention of harmful emissions in the future. The question of technology and resource transfer, which is at the core of the conflict, remains to be settled.

Mitigating other types of local and regional pollution is sometimes quite burdensome in the current financial climate in the developing economies of the Asian and Pacific region. Fuel switching from oil to natural gas as a policy option, as well as on economic and environmental merits, is taking place in many countries of the region. However, at the same time the use of coal is also increasing. Clean coal technology has to be adopted if environmental impacts are to be minimized.

B. Energy efficiency improvement options

To mitigate environmental degradation, the most viable option for the developing countries would be to burn fossil fuel more efficiently and to increase the efficiency of the end-use equipment so that the energy intensity can be reduced. This issue is discussed in greater detail in a separate paper "Prospects for Energy

---

5 Toufiq A. Siddiqi, *Interpreting Sustainable Development in the Energy Context* (Honolulu, Hawaii, East-West Centre).

Sector Implementation of Agenda 21: The Role of Energy Efficiency", of this publication.

VI. CONCLUSIONS AND RECOMMENDATIONS

A. Sustainable development and utilization of energy

It is evident that energy demand in the region, particularly in developing countries, will continue to grow at a high rate. The main challenge for the region will be how to manage the growing demand effectively within financial, technical and environmental constraints. A shift towards an alternative energy development pathway in the area of resource options, improved efficiency and clean fuel technology, would be a way to move forwards to sustainable development.

Energy policies need to be reviewed and adjusted in the light of the concern about the environment and the sustainability of energy.

Integration of the environment in all phases of planning, development and management of the energy sector is a necessary initial step towards sustainable development. Although there are some differing opinions between the developed and developing economies on the issue of global warming, there is no difference in opinion on the issues of local and regional pollution (such as sulphur dioxide and nitrogen emissions, particulate emissions and waste disposal). According to the situation in each country, a priority action plan can be developed to mitigate environmental degradation. A standard should be set to monitor, prevent and control pollution.

The implementation of sustainable development policy involves additional costs in the energy sector. A mechanism should be developed to internalize these costs in the pricing of energy. An appropriate pricing signal is needed for attracting private sector participation in the capital investment. Progress in the implementation of environmental policy in the energy sector of developing countries will also be dependent on the transfer of technology and the access to additional funds.

B. Regional cooperation

Although countries in the ESCAP region are diverse in their level of development, they share many issues in common with regard to energy development. They could benefit from the sharing of information and experience. To facilitate such exchanges, the regional Working Group on Energy and Environment Planning and five other working groups covering various subsectors of energy were established in the Asian countries of the ESCAP region within the framework of the former UNDP/ESCAP Regional Energy Development Programme (REDP). They are being strengthened under the UNDP/ESCAP intercountry programme for Asian cooperation on energy and environment.

Regional cooperation may take many forms, such as a regional forum to discuss energy strategies and policies, an Asian energy efficiency programme for the twenty-first century, a TCDC programme on new and renewable sources of energy and studies on regional collaboration on energy and power systems, including all traditional supply sources (oil, gas, coal, etc.) based on an environmentally sustainable utilization pattern.

CHAPTER II

ENVIRONMENTAL PROBLEMS AND POLLUTION CONTROL STRATEGIES RELATED TO PRODUCTION AND USE OF ENERGY IN CHINA[*]

1. ENERGY PRODUCTION AND CONSUMPTION IN CHINA

Energy provides the basic power for national economy and people's livelihood. However, production and use of energy could also cause serious damage to environment. Therefore, it is of vital importance to control pollution caused by production and use of energy.

China is a developing country. The energy production and consumption is increasing rapidly and the structure of energy is changing slowly year by year in order to satisfy with its social and economic development. In 1990, the total energy production and the total energy consumption reached 1039MTCE and 980MTCE respectively. Of the total production, coal accounted for 74.23 per cent, oil 19.01 per cent, natural gas 1.96 per cent, hydropower 4.80 per cent. Of the total consumption, coal accounted for 75.6 per cent, oil 17.0 per cent, natural gas 2.1 per cent, hydropower 5.3 per cent. Coal is the major form of energy in China.

Of coal consumption structure in China, 25.6 per cent of the total coal consumption is used for generation electricity. The proportion of civil use is 21.6 per cent. In addition, it is revealed in some reference that 84.3 per cent of civil fuel in some cities is provided with coal.

It is a long term energy policy in China to use coal as the main part in energy composition. It is estimated that the total coal consumption will reach 1400 MT.

2. ENVIRONMENTAL PROBLEMS RELATED TO PRODUCTION AND CONSUMPTION

Because coal is main energy resources, a lot of air pollutants and warming gases mainly result from the production and use of coal in China.

(1) Sulphur dioxide emission and acid deposition

The average sulphur content of coal is 1.25-1.35 per cent in China. Southwest China is the region where the sulphur content of coal is higher. Sulphur content of coal in Northeast and North China is lower.

It is reported that the national total amount of SO_2 emission was 15 million tons in which 90 per cent of SO_2 comes from coal combustion in 1990.

The monitoring information of air quality in 32 cities in 1990 showed that the average SO_2 concentration in 16 northern cities is 0.109 mg/m^3, 2.5-1.8 times the WHO standard (0.040-0.060 mg/m^3); the average SO_2 concentration in 16 southern cities is 0.102 mg/m^3, 2.5-1.7 times the WHO standard.

Pollution of acid deposition has happened in some regions in China. The regions suffered from acid rain pollution are Southern, Southeast and Southwest of China. Now the tendency is becoming more serious.

(2) Dust Emission and Pollution

In China, the total amount of dust emission was 21 million tons in which 63 per cent is from coal combustion. The average particulate concentration in 16 northern cities is 0.502 mg/m^3 which is as 8.4-5.8 times the WHO standard (0.060-0.090 mg/m^3), the average particulate concentration in 16 southern cities is 0.267 mg/m^3, 4.5-3 times the WHO standard.

(3) Carbon Dioxide Emission

Consumption of fossil fuels is the major emission source of CO_2 in China. It was estimated that the emission of CO_2 may be about 600 million tons of carbon a year. The 83.2 per cent of total emission in China is from coal consumption, 15.5 per cent from oil use.

(4) Nitrous Oxide Emission

It is very difficult to estimate N_2O emission. The sources of N_2O is generated mainly from biological

[*] Presented by Chai Fahe, Chinese Research Academy of Environmental Sciences, Beijing, at the Expert Group Meeting Preparatory to the First Session of the Committee on Environment and Sustainable Development, Bangkok, 30 September – 2 October 1993.

activities and burning of biomass and fossil fuels. It is pointed out by Zhuang Yahui et al that the total N_2O emission is about 0.34 million tons a year in China. 89.6 per cent of the total emission is from consumption of fossil energy such as coal, oil and natural gas.

(5)   Methane Emission

Methane emission mainly comes from rice paddies, coal mine, leakage of natural gas and biogas pond, ruminant, landfill and biomass burning. It is estimated that methane emission from energy production such as coal mine and natural gas production is about 8 million tons of carbon which is less than methane emission from rice paddies 33 million tons.

(6)   Nitrogen Oxide Emission and Pollution

NOx has 6 per cent contribution to the global warming. NOx emission in China is relatively low (0.85 million tons a year). It mainly comes from coal combustion (70 per cent). NOx emission (29 per cent) from cars is very little compared with developed countries because the number of cars in China is very small.

The average NOx concentration in 16 northern cities is 0.008-0.120 mg/m^3. The average NOx concentration in 16 southern cities is 0.0009-0.110 mg/m^3. The annual daily average NOx concentration exceeds national standard only in Zhengzhou city and Guangzhou city.

3. STRATEGIES FOR ENERGY DEVELOPMENT AND POLLUTION CONTROL

China government has paid great attention to solving environmental problems related to energy production and consumption. In order to control pollution emissions which impact regional evironment and have a contribution to global climate warming, the government has made a series of policies and detailed plans for reducing emissions of greenhouse gas and other air pollutants. In China, energy development and energy saving are equally important. The national energy development will be still based on coal. In addition, main ideas of national energy policy is to make strenuous efforts to develop electric power, especially hydropower, nuclear power, oil and natural gas, to encourage power saving, oil saving and coal saving, to spread co-generation of heat and power, to utilize waste heat from power station and other boiler houses, to enhance efficiency of energy use, to reduce emission of pollutants related to production and consumption, to

improve air quality, to decrease contribution to climate warming.

Solving environmental problems related to the production and consumption of energy is a complex system engineering. The environmental problems in primary energy production are relatively simple. It is clear that the essential reasons of environmental problems caused by energy use in China can be ascribed to the low efficiency of energy use, the irrational structure of energy and low level of harnessing tail gas. In my opinion, the following measures must be taken to solve environmental problems related to the production and consumption of energy in China.

(1)   Reducing Emission of Pollutants from Primary Energy Production

The natural gas and methane are both valuable energy resources. Its emissions to atmospheric environment not only squander resources, but also pollute atmospheric environment and result in greenhouse effect. Therefore, it is important to prevent leakage of natural gas and collect and utilize methane from coal mines.

(2)   Raising Efficiency of Energy Use

It is noticed that the efficiency of energy in China is about 30 per cent which is far lower than that in developed countries. It is possible to increase the efficiency and decrease the quantity of energy use so as to reduce the emission of air pollutants. If the efficiency of energy use is increased from 30 per cent to 60 per cent, the emission of carbon dioxide and other air pollutants will be decreased by 50 per cent.

Most of industrial boilers in China burn coal directly. Some old boilers which has been used for several decades are still operating. Improving the efficiency of burning and heating of industrial boilers and substituting for the old boilers with advanced boilers is one of the most important measures to decrease the amount of coal use in China.

Spreading energy saving technology such as controlling operation of boilers by means of microcomputers is also helpful to improve the efficiency of energy use. It was shown that controlling operation of boiler by means of microcomputer can save 5-7 per cent of coal.

(3)   Improving the Quality of Energy

Because the calorific value and the size of coal cannot come up to the requirement, coal cannot be burned enough and the efficiency of heat is decreased in most of industrial boilers except those in power stations. If the raw coal is washed, desulphurized, crushed, sieved and mixed with different kinds of coal to meet the demand of burning in boilers and to remove harmful impurities contained in the raw coal before it is used, the heat efficiency of industrial boiler will be greatly increased and the emission of air pollutants will be obviously decreased.

Another way which improves the quality of energy is to change the form of energy. For example, coal may be transformed into gas and/or liquid energy by gasification and/or liquefaction of coal or moulded into hard cubes or balls of the same size.

(4) Changing Structure of Energy Consumption

Directly burning coal is one of the most important reasons which leads to air pollution and emission of greenhouse gas. An effective measure would be to change the structure of energy. That is the proportion of clean energy should be increased and the proportion of coal consumption should be decreased.

a.   Transform more coal in electric power production.

b.   Go all out for hydro-electricity construction. Efforts should be concentrated on the construction of a large number of water conservation energy and hydro-electricity projects, such as 3 Gorges project. In the countryside, if there is potential to develop small hydro-electricity, Government should encourage and help to build small-sized hydro-power station to replace part of firewood and coal.

c.   Develop nuclear power.

d.   Try to use natural energy such as solar, wind, biomass, and tidal power.

e.   Spread technology on biogas in the vast rural areas.

(5) Enhancing the Level in Treating Equipments of Tail Gas

The level in treating the tail gas from boiler is lower because of the reasons in economy and technology in China. Only there are treating dust equipments to be installed in the industrial boilers. There are very few boilers with large capacity to be installed with equipments which control sulphur dioxide emission. For carbon dioxide, there are no controlling measures.

China have matured experiences and foundation in removing dust from boilers. For sulphur dioxide emissions, there is matured treating technique in China, but it is too expensive. The national EPA has begun to try collecting fees from emitters of sulphur dioxide to accumulate foundation and develop new kinds of sulphur dioxide emission control technology.

The Government of China attaches great importance to control carbon dioxide emission. But, I think developed countries should do first and help developing countries in economy and technology in the field because they have emitted a lot of carbon dioxide since industrial revolution; another reason is that developed countries are rich in economy and have technology on carbon dioxide control. The developed countries have an obligation to transfer foundation and technology to the developing countries in order to build capacity of carbon dioxide control.

(6) Reducing Emission of Pollutants by Means of Economy

It is clear that environment also is a kind of resources. The polluter-pays principle should be adopted to limit behaviour of emitters and allocate the costs of pollution control.

In China, there are two kinds of economic methods in environmental management, i.e, 1) restrictive method including emission charge, fuel tax and penalty and 2) encouraging method including tax exemption, emission trade and government supporting policy.

The restrictive method is to impose a fine to the enterprises and institutions where emissions of pollutants exceeds national or local standards on emission of pollutants. The method gives polluters an external economic pressure to improve internal production and environment management, save energy, reduce emissions of pollutants to achieve the goal of protecting environment and saving energy.

The encouraging method is to promote the enterprises to control pollution and save energy by financial support. In addition, the emission trade of pollutants among emitters should be allowed if the trade can reduce total emission of pollutants, improve the environment quality and save energy.

(7) Strengthen Environment and Energy Management

There are environmental management regulations which consists of eight regulations, i.e, the Three Simultaneous Regulation (TSR), Environmental Impact

Assessment (EIA), the Regulation of Punishment with Fines for Pollutants discharge (RPF), the Target Responsibility System for environmental protection (TRS), the Quantative Appraisal System on comprehensive urban environmental control (QAS), the Centralized Pollution Control System (CPCS), the Regulation for Pollutant Discharge and Pollutant Discharge Permit System (RPDPS) and the System of Setting Deadlines for Pollution Source (SSD). Each of the regulations is related to control pollutants from production and consumption of energy.

The development of Energy industry also takes the environmental protection into account. The national energy development strategies include changing structure, increasing the proportion of clean energy, mainly developing electric power, especially hydro-power, cleaning coal, limiting directly burning coal, co-production of power and heat, supplying gas in urban area, falling backward boilers into disuse, using high efficiency and energy saving boilers, saving energy, etc. These strategies are not only advantageous to development of energy industry, but also favourable to environmental protection.

CHAPTER III

ENERGY DEVELOPMENT AND ENVIRONMENT MANAGEMENT*

Environmental concerns of energy use have been debated for some time now and have assumed more significant proportions in recent years with definite information being now available about their effects on global warming climatic changes. The enormity of the perceived impacts is brought out starkly by the level of international effort that is currently going into formulating strategies to contain the build up of greenhouse and other emissions. While a number of developed countries have reached a stage when they can afford to reduce the overall consumption of energy without adversely affecting their progress, developing countries including India on the other hand find it extremely difficult to contain growth in rates of energy use.

ENERGY SCENARIO IN INDIA

In a developing country like India where per capita consumpiton of energy is quite low (equivalent to 273 kg of oil equivalent in year 1987 as against more than 4,500 kgoe in the OECD countries), the demand for energy is bound to grow to facilitate all round socio-economic development. The past trend in per capita energy consumpiton during the quarter of a century between 1952 and 1987 indicates a tendency towards more intensive overall use of electricity and commercial energy in the Indian economy (Table I).

Table I: Per Capita Final Energy Consumption

| | --Energy Consumption-- | | Electricity | % Share of Commercial Energy |
| | Total | Commercial | | |
	(KGOE)	(KGOE)	(KWH)	
1953	224.4	58.3	21	26.0
1960	244.7	73.4	38	30.0
1970	246.0	85.7	90	34.8
1980	262.9	107.1	132	40.7
1987	273.4	130.0	203	47.5

Source: Sectoral Energy Demand in India, Regional Energy Development Programme, UN-ESCAP Report No.RAS/86/136, 1991

---

* Presented by Ajay Dua, Director, Energy Management Centre and Joint Secretary, Ministry of Power Government of India at the Expert Group Meeting Preparatory to the First Session of the Committee on Environment and Sustainable Development, Bangkok, 30 September – 2 October 1993.

The per capita final energy consumption has increased from 224.4 kgoe in 1953 to 273.4 kgoe in 1987 at an implicit average annual growth rate of 0.6 per cent i.e. the per capita primary energy consumption in 1987 was 1.4 times that in 1953. The share of commercial energy consumed rose from 26 per cent in 1953 to 47.5 per cent in 1987 at an average annual rate of 2.4 per cent. In case of electricity, the per capita consumption growth has been very much higher and it grew from 21 kWh to 203 kWh at 6.9 per cent per annum.

The growth in energy consumption in developing countries tends to be correlated to the growth in GDP and India is no exception. In the case of commercial energy, the elasticities with reference to GDP continue to be more than unity (1.25 between 1970 and 1987) for the reason that the growth in commercial energy consumption is not only on account of the increase in the activity level in the economy but also due to replacement of traditional fuels by commercial fuels. With the rapid urbanization and noticeable industrialisation taking place in India, the traditional sources of energy mainly bio-gas, fuel wood, crop residue and animal waste which account for around 50 per cent of the total energy consumed are being substituted by commercial energy sources. It is estimated that the total requirement of energy in the economy could be 2.2 times higher in 2009 than in 1980 and the requirement of commercial energy will grow by 6.3 per cent annually taking its share in total energy from 43 per cent in 1986 to 80.5 per cent in 2009.

The two figures given below show the primary sources of energy in India and the share of various sectors in it.

In the last four decades, there have been noticeable changes in the pattern of use of commercial energy. The share of agriculture has increased steadily from 1.6 per cent in 1953 to 9.0 per cent in 1989 while the share of transport has declined during this period from 43.6 per cent to 24.5 per cent. The share of industry and the domestic sector however has grown substantially (of industry from 39.8 per cent in 1953 to 51.5 per cent in 1980 and then decreasing to 50.4 per cent in 1989 and of the household sector from 4.4 per cent in 1953 to 13.8 per cent in 1989).

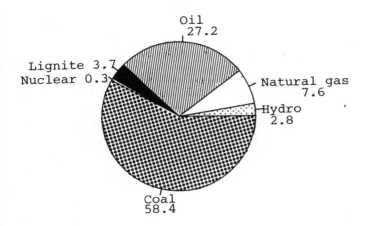

Figure I. Sectoral Share of Primary Source

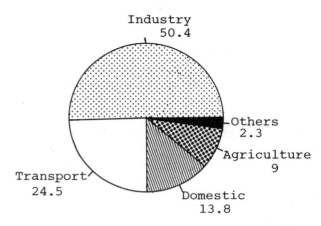

Figure II. Consumption of Commercial Energy

The sectoral commercial energy intensities for different years are indicated in Table II :

Table II: Sectoral Commercial Energy Intensities
(KTOE/ Rs Billion)

Sector	1953	1960	1970	1980	1987
Industry	135.6	138.3	144.0	159.1	115.9
Agriculture	1.4	1.8	3.4	9.5	15.7
Transport	908.5	802.9	479.3	333.0	304.4
Household	4.6	5.8	8.2	8.2	10.2

Source: Sectoral Energy Demand in India, UN-ESCAP REDP (RAS/86/136), 1991

The primary enery intensity of the economy w.r.t. GDP (at 1980 prices) has declined from 190.6 ktoe/Rs billion in 1953 to 158.5 ktoe/ Rs billion in 1987 mainly on account of shift to more efficient commercial fuels. The intensity in economy of commercial energy has increased from 43.7 to 59.8 ktoe/ Rs billion during this period. (1 US$ = approx Rs 31)

The energy-intensity in the case of industrial sector in India increased from 136 ktoe/Rs billion in 1953 to 159 ktoe/Rs billion in 1980 and declined thereafter to 116 ktoe/ Rs billion in 1987. This was perhaps on account of a greater share of high value added industries, viz. electronics, petro-chemicals, etc. and other structural changes in the industrial sector over the years. In the case of the transport sector, the energy intensity has declined steadily from 909 ktoe/Rs billion in 1953 to 304 ktoe/ Rs billion in 1987. The decline was sharp in 1980 whereafter it somewhat slowed down. This occurred mainly due to a shift away from steam traction to more efficient diesel and electric tractions. On

the other hand, the energy intensity in agriculture and households sectors increased over this period. In the household sector, the commercial energy intensity has increased on account of urbanisation and general improvements in the standard of living which have resulted in a larger proportion of population using commercial sources of energy in place of traditional sources. In agriculture sector, the higher energy intensity has been on account of expansion of ground water irrigation and mechanisation.

GROWTH OF POWER SECTOR

India's power generation industry has grown rapidly over the last four decades after it was nationalised in 1948. The installed generation capacity has increased fifty three times from 1,362 MW in 1947 to over 72,300 MW in 1992-93 and compares in size with the power generation industry of UK, Germany and France. Supply of electricity however continues to lag behind the demand for it.

Demand for electrical power in India which has increased by 8 per cent annually over the last two decades would continue to grow rapidly in the future. It is estimated that during the VIII Five Year Plan i.e. between 1991-92 and 1996-97, the demand for electricity would grow by about 10 per cent annually as against an expected growth of 5.5 per cent in GDP. Such a trend in growth in consumption is likely to continue even in the decade beyond the VIII Plan.

During the current Development Plan, it is envisaged to augment the generation capacity by 30,537.5 MW. However, even with this order of capacity-addition, the power availability at the end of the plan-period vis-a-vis the demand for it, would not be any better than at the beginning of the Plan when there was an 8 per cent overall electric-energy shortage and a peak time deficit of about 20 per cent. Serious doubts are already being raised about India's ability to increase the generation capacity to the extent indicated in the Plan since the resources provided for it are not commensurate with the current costs of creating such capacities. India's power sector had in the last two decades received substantial external financial assistance including from the erstwhile Soviet Union, a source which has abruptly disappeared without any reliable alternative source emerging. The current trends in supply and demand for power in fact portend a future which is not too bright both metaphorically and literally speaking.

The costs of producing power are increasing in terms of the capital required as well as the operating costs. The investment required for creating new capacities is on the rise as new sources of fuel are being used for power-production e.g. gas and nuclear. Also increasingly expensive equipment for pollution control and environmental protection is being required to be installed in new power stations. The operating costs of generating power are increasing as coal, the primary fuel for producing electrical power in India is now being required to be transported over much longer distances than before and the quality of coal is steadily deteriorating e.g. power plants in India nowadays often receive coal which has 40 per cent ash content. Also though India annually produces about 25 million tonnes of oil, it imports at least as much to meet the current demand and the costs of petroleum-products and natural gas have risen substantially in the last two decades.

Power Sector and Environment

Total installed generating capacity in the power utilities at the beginning of the 8th Plan i.e. 31st March, 1992 was 69,082 MW comprising of 48,108 MW Thermal, 19,189 MW Hydro and 1,785 MW Nuclear. The total generation of electricity during the year 1991-92 was 286.7 billion units and there was energy shortage of 8.5 per cent on all-India basis and peak power shortage of 17.7 per cent. Region-wise the peak load shortages varied from 32.4 per cent in the Eastern Region to 7.6 per cent in the Western Region and the energy shortage varied from 21.6 per cent in the Eastern Region to 8.3 per cent in the North-Eastern Region. The demand for power in India is increasing at around 9 per cent annually while the supply increase in the last 15 years has been around 6 per cent. Consequently, the energy shortage and the peak shortage has been growing and it is anticipated that even if 30,538 MW of capacity as provided in the VIII Five Year Plan is added, the overall energy shortage and peak time shortage would

not be reduced by 1996-97. The share of hydro generation capacity which plays an important role in meeting the peak demands has declined from 34 per cent at the end of the VI Plan to 29 per cent by the end of the VII Plan and further to 27.8 per cent at the end of the VIII Plan. The share of the Central Government owned generating companies in the total installed capacities has been steadily increasing from 16 per cent at the end of the VI Plan to 26.1 per cent at the end of the VIII Plan. The share of large coal pithead power stations has increased progressively from 10 per cent at the end of the VI Plan to 25 per cent at the end of the VII Plan.

Environmental concerns associated with production of electricity are emerging as significant issues and there is a concern about the potential adverse effect of power generation on land, water and air resources depending on the type of power plant used. Coal continues to be the main stay of power generation in India and this placed an enormous burden on the power sector for contributing to emissions. Even hydro electric projects are a matter of concern since they can submerge large tracts of land, cause soil erosion, deforestation and disturb adequate life. Extra high voltage systems are also known to cause audible notes in their vicinity and induce electromagnetic fields which can have adverse effects on habitation and the surrounding flora and fauna. Nuclear power plants, radio active waste management and potential risks of global climate change induced by increasing atmospheric concentration of carbon dioxide and other green house gases are matters increasingly bothering the Indian policy makers.

Agenda 21 of the U.N. Conference on Environment and Development (UNCED) held in Rio de Janeiro in June 1992 which is now guiding the policy makers the world over attempts a fairly comprehensive approach recommending national planning to integrate energy, environment and economic policy in a sustainable frame work. The internalization of environmental costs through economic (incentive based) and regulatory (fiats) measures is also urged as a planning goal. It has been noted in this regard that economists generally prefer incentive based instruments which, in effect price the resource to the user, to fiats. The former under a set of assumptions will permit a given level of environmental quality to be reached at the least resource cost. Several incentive based instruments such as pollution taxes, tradeable permits, may also be designed to raise sufficient revenues for funding sustainable development measures or meeting equity concerns. However Agenda 21 does not urge the adoption of (economic) regulatory instruments in a multilateral frame work i.e. with States as the regulated agents.

One important set of policies which may have significant complementarities with economic efficiency goals is increased energy efficiency and conservation. The reduced use of primary energy, in particular fossil fuels through increased energy efficiency will reduce polluting emissions and discharges, and thus promote sustainability. Accordingly, Agenda 21 also proposes the development and use of energy efficient technologies, stating goals for energy efficiency and technology transfers to developing countries. Of course under the climate convention, the responsibilities of developing countries for implementing national abatement strategies is conditional not only on the fact of such technology transfers but also on the 'agreed full incremental costs' ('agreed' under norms to be decided) being met by industrialised countries. Questions relating to the appropriate depth of technology transfer would need to be addressed in future negotiations. Several energy sources are identified as 'sustainable' and therefore as meriting increased research. These include solar, wind, geothermal, hydro power and bio-mass (including wood). All of these, of course, have a range of environmental impacts.[1]

COAL SECTOR – ENVIRONMENT AND DEVELOPMENT ISSUES

Indian Coal Industry

Coal is India's most abundant source of energy currently meeting over 58 per cent of the country's primary energy requirement and will continue to remain the main source of energy in the foreseeable future. Sustained development of all facets of the coal industry is, therefore, imperative for overall economic development of the country. India's known coal resources have been assessed to be about 194 billion tonnes and are enough to sustain an annual production level of around 550 million tonnes well beyond the 21st century.

Immediately after the nationalization of the coal industry, massive development programme was undertaken in reorganizing the existing mines and implementing new projects. Since the nationalization, development of 405 coal projects each costing Rs. 20 million and above has been taken up (US$ 1=approx Rs. 31). The total sanctioned investment on these projects is of the order of Rs.15 billion for ultimate production capacity of around 310 million tonnes per annum. Of these construction of over 254 projects has been completed and these are producing coal.

An investment to the tune of about Rs.15 billion covering new projects, existing mines and other

infrastructure has been made in the nationalized coal sector during the period 1973 to 1992. The coal production which was about 77 million tonnes in 1972-73 rose to a level of about 238 million tonnes in 1992-93, registering an annual growth rate of 6.1 per cent. Power utility sector is the single largest consumer of coal. Next to electricity sector, steel is the second largest consumer of coal and cement industry is in the third position. In 1992-93, the total consumption of coal was 243.52 million tonnes, of which 61 per cent was consumed for electricity generation, 13 per cent for steel making and 4 per cent in cement industry. The balance 22 per cent was consumed by other industries including captive generation of power. The coal based installed capacity which was around 8,000 MW in the year 1973, when the coal sector was brought under the State control, has grown to a level of over 46,000 MW at present which is 65 per cent of the total installed capacity. The coal based installed capacity is expected to go up to a level of 77,000 MW out of a total capacity of 136,000 MW projected for the year 2001-02. The power sector will therefore remain the largest coal consumer in future.

Growth in consumption of coal in different sectors is given in the Table 1 given below :

Table 1

	1979-80	1984-85	1989-90	1992-93	(in million tonnes) Annual Growth over 1979-80(%)
Steel	22.52	25.27	28.37	32.13	2.8
Power (Utilities)	34.86	64.35	115.12	149.25	12.0
Railways	12.53	9.53	5.73	3.22	–10.0
Cement	4.53	7.11	8.74	10.71	6.8
Fertilizer	2.11	3.89	3.97	4.53	6.0
Soft Coke/SSF	3.38	2.24	1.30	0.63	–12.1
Other incl.captive power	28.85	31.59	38.68	43.05	3.4
	108.78	143.98	201.91	243.52	6.4

The indigenous coal production is, by and large, adequate to meet the overall demand of various consumers except the steel plants for which about 6 million tonnes of low ash coking coal is annually imported partly to meet the gap in domestic availability and partly for technological requirement of quality improvement. Action has been initiated to gradually reduce the import of coking coal in the Eighth and subsequent Plan periods by increasing domestic availability for the steel plants.

Growth Plan of Coal Industry

For meeting the projected coal demand of 318 million tonnes the terminal year of VIII Five Year plan, a production programme of 308 million tonnes has been drawn which together with the drawal from the accumulated pit-head stocks would be adequate to meet the projected coal demand. Planning preparedness for achieving the targets of VIII Five Year Plan has already been ensured and the implementation of sanctioned and new projects together with coal production from the existing mines would be adequate to meet the coal demand of the VIII Five Year Plan period. The contribution pattern from different group of mines is as under:

Table 2

	(in million tonnes)	
	1992-93	1996-97
Existing Mines/ Completed Projects	170.00	133.35
On-going Projects	63.50	157.06
New Projects	–	12.59
	233.50	303.00

* Another 5.00 mt will be the contribution from the captive mines of TISCO/IISCO/DVC.

By the terminal year of IX Plan i.e. 2001-02 the demand of coal is expected to grow to a level of about 396 million tonnes. Planning preparedness to meet this demand has already been completed. Projects for an ultimate capacity of about 70 million tonnes are under appraisal and scrutiny and are likely to be approved during the VIII Five Year Plan. To facilitate expeditious implementation of new projects, 30 advance action proposals covering initial activities like acquisition of land, rehabilitation of persons likely to be displaced and creation of initial infrastructure have already been sanctioned involving a capital outlay of Rs. 1,850 million. These proposals are geared to contribute the bulk of the incremental production during the IX Five Year Plan.

Environmental Concerns in coal mining

The complete Coal Chain comprises of mining, transportation and end-use. The environmental impact of coal mining and that of the end use are generally site specific and identifiable. The environmental impacts of transportation of coal are more spread out, less identifiable and their quantification becomes difficult due to secondary pollution caused by the fuel (generally diesel oil) used by the transport system. The quality of coal has significant impact on the environment particularly at the consumption end namely the thermal power plant or cement factory etc. Air pollution is the major concern which needs to be addressed. The secondary pollution caused by the mode of transportation is related to the coal bulk handled by the system. In this context, beneficiation or cleaning of high ash coals is particularly relevant to India.

During the process of mining, beneficiation and handling of coal, major environmental concerns which are to be addressed are restoration and reclamation of land, afforestation, relocation of people affected by mining activity, air pollution and water pollution including disturbance to the ground water. All these aspects are carefully attended to while preparing an Environmental Management Plan which incorporates assessment and quantification of impacts as well as abatement and restoration measures for dealing with the various environmental concerns. Rejects and middlings which are by-products of coal washing are required to be disposed in an environmentally friendly manner. In case of a two-product washery, the rejects contain some burnable coal with gross calorific value of around 1600-1,700 kcal/kg. These can be used in fluidised bed boilers for power/steam generation. In a three-product washery also the rejects have fuel value which can be recovered through FBC boilers. In case the fuel value of rejects is very low, these are required to be disposed in the mined out areas after taking suitable precautions against spontaneous heating. Middlings having heat value in the range of 3,000-4,000 kcal/kg are supplied to the power stations located near the coal fields.

Coal mining activities particularly the open cast mining generally require a large area of land. Open cast mining have adverse effects due to degradation of land, air, water and noise pollution. Due to lack of locational options, damage to forest areas and shifting of people living within the mining blocks become inevitable. Environmental problems in India have further been compounded by unscientific mining in the past leading to land degradation and extensive damage to the overall environment in the mining areas. About 18,000 hectares of the land has been rendered derelict due to coal mining operations in the past resulting in fires, subsidences, open excavations and waste dumps. If these degraded areas are to be completely restored then the costs are going to be enormous. Therefore, a balance has to be struck by reclaiming part of the affected areas to make them fit for the afforestation and the remaining areas can be converted into water bodies for irrigation as well as for pisciculture.

Currently the coal industry is rendering about 500 hectares of land biologically unproductive every year mainly because of the emphasis on open cast mining. It is anticipated that land degradation will rise to about 1,500 hectares a year by the end of the IX Plan period. This degraded land needs proper rehabilitation. Another factor is the effect of mining in forest areas. Fairly large tracts of forest land in North Karanpura, Ib valley, Talcher, Singrauli, Korba, Wardha and Godavari valley coal fields will be affected as a result of large scale open cast mining in these areas. Due care will have to be taken to ensure adequate compensatory afforestation work. In India because of high population density factor, land degradation is very high and therefore reclamation of land assumes great importance. In 1984-85 about 174 million hectares of land was being affected by degradation and soil erosion of which about 40 million hectares are reported to have been recovered due to various reclamation and restoration programmes undertaken by the Government.

With bulk of high ash coals being used for generation of electricity, the problem of fly ash is quite acute. Presently about 50 million tonnes of fly ash are being generated annually and are likely to go up to 85 million tonnes by the year 2001-02 if unwashed coals are continued to be used in all power plants irrespective of distance from the supply sources. About 2 per cent of the fly ash is being gainfully utilized at present. Some of the uses of fly-ash generated from coal fired power plants are :

- concrete, concrete products manufacture
- road making
- cement manufacture
- bricks, blocks and cellular concrete
- mixing with cement in mortar
- land fills
- extraction of alumina and rare earths

In view of the environmental degradation caused on account of certain developmental activities and also considering the necessity to protect and improve quality of environment by controlling air, water pollution and soil along with biotic pressure on natural resources, prior environmental clearance from the Government is required under the Environmental Protection Rules (1981) before a new project or expansion of an existing project is taken up. For the purpose of institutionalizing the environmental clearance at the level of Central or State Govt, various projects have been separately listed. In case of specified projects like mining, pit-head thermal power stations; hydro electric projects and ports/ harbours prior environmental clearance of the location will also be required under the new dispensation. The institutional mechanism for examination and scrutiny of EIAs and EMPs comprises of multi-disciplinary expert committee. There is also a provision in the rules for effective monitoring of implementation of environmental protection measures and the conditions subject to which environmental clearance has been given. The project authorities are required to submit half yearly reports to the Ministry of Environment and Forest.[2]

Clean Coal Technologies in Power Sector

The present thermal efficiency attainable with pulverised coal firing system in Indian power stations is 33 per cent on an average basis and 36 per cent on peak basis with little potential for further improvement. Clean Coal Technologies which are more efficient even with high ash Indian coals could improve the productivity of existing thermal power plants. The available option of technology consist of gasification, pressurised fluidised bed combustion, coal beneficiation for slurry formulation and its combustion. Government of India have set up a Technology Mission and it proposes to implement through demonstration plant in the following :

- Integrated gasification combined cycle (IGCC);
- Pressurised fluidised bed combustion (PFBC); and
- Coal Beneficiation and slurry combustion (CBSC).

The time frame for implementing these programmes is 5-6 years and total estimated fund requirement is around Rs.4.87 billion.

Till the commercialisation of these new Clean Coal Technologies, beneficiation of coal for improving its quality and consistency by reducing ash content will improve the pulverised coal fired power plants' performance and reduce pollution. A few beneficiation plants are already under various stages of construction. As a medium term effort, beneficiation of coal is called for and it would not only reduce carbon and sulphur dioxide emission to a very large extent but also improve overall plant performance and environment. Considering the high ash content of Indian coals and erosive nature of ash, fluidised bed combustion system has been identified as a short term solution to efficient utilisation of Indian coal for industry as well as small to medium cogeneration plants.

HYDRO POWER & ENVIRONMENT

Hydro power has the potential for submerging large tracts of land and cause soil erosion, deforestation, disturb aquatic life and reservoir induced seismicity. All these issues need to be carefully addressed while designing projects including run-of-the-river types of the scheme which have negligible storage and generally need to be preferred to storage type schemes which cause large submergence. At present storage schemes form about 40 per cent of the total number or schemes identified to develop India's entire hydro power potential of about 200,000 MW including from non-conventional sources.

Setting up of hydro projects needs to be accompanied by detailed catchment treatment plan to check the soil erosion, control the siltation in the reservoir and to activate the biological processes. Such a requirement is now being insisted upon in India with respect to all hydro projects and upgraded areas in the catchment are required to be treated by suitable soil conservation measures and check dams are required to be built across the tributaries and small rivulets prone to silting. Similarly the loss of forest due to execution of water resources projects is now required to be compensated by raising afforestation over an equivalent non-forest area or double the area of degraded forest in case of non-availability of non-forest land which is binding on the projects in terms of the Forest Conservation Act of 1980. Submergence of flora and fauna and possible extinction of some rare species are possible barriers to fish life in hydro projects. Therefore, with respect to new projects, environmental impact studies are required to be undertaken with the objective of ascertaining any possible threat to rare species found in and around the project area, impact on fishes of economic value and the magnitude of import of fish food organisms. Thereafter a suitable design and placement of fish ladder has been made compulsory if necessary and feasible. Other methods like fish passage, fish way or setting up of fish farms are also undertaken.

The impact of hydro power projects on green house is mostly attributable to deforestation if it occurs as a result. But if compensatory afforestation is insisted upon then this adverse effect can be minimised.

There is some ambiguity about the apprehension or impounding of water induced seismicity. Seismological observations established at Bhakra, Pong and Ram Ganga dams in the Himalayan terrain have not registered any increase in seismicity due to impounding of water. Earthquake engineering experts feel that precautions need to be taken by resorting to slow the rate of reservoir filling even though there is no definite proof that impounding of water leads to a greater possibility of occurrence of earthquakes.

Certainly there is need to ensure that whenever any hydro project, large or small is taken up, the original residents are resettled in a manner which improves their overall quality of life rather than adversely affect it. Consequently nowadays comprehensive rehabilitation packages are required to be drawn up for project oustees keeping in mind the peoples' own priorities, social customs and their economic needs and preservation to the extent possible of the social characteristics of their life style. Apart from providing the basic amenities at the resettlement sites, social, cultural and religious, amenities are also required to be provided by the project authorities and an agreed rehabilitation package is required to be drawn up. These include provision of land in lieu of the land submerged or compensation for land or property at the prevailing rates besides employment to members of families and rehabilitation-grants to the landless or those being rendered houseless.

Since impounding of water has the potential of affecting the life of bacteria, protozoa and other micro organisms in the riverain eco system there is always a danger of spread of water related diseases such as malaria, filariasis, diarrhoea and schistosomiasis. Therefore there is need to mitigate these dangers by adopting improved design parameters, proper drainage, regular flushing, lining of canals and use of insecticides etc. This problem is not envisaged in case of run-of-the-river schemes where pondage is limited to a few hours of plant running and the water is continuously replenished.[3]

PETROLEUM SECTOR – DEVELOPMENT AND ENVIRONMENTAL ISSUES

Oil production which was only 0.25 million tonnes in 1947 reached 1 million tonne mark in India in 1962 and hovered between 7 and 8 million tonne from the late 1960s to the mid-1970s. However there was a quantum jump in oil production in the beginning of 1980s due to full exploitation of Bombay High Offshore and production went up to about 29 million tonnes by 1984-85 and was 34 million tonnes at the end of the VII Plan (1989-90). There has been a decline in the oil production since then due to planned containment of Bombay High production and oil production in 1992-93 was only about 25.3 million tonnes. About 60 per cent of total oil production in 1992-93 was from offshore fields.

Natural Gas production has been rising steadily since the beginning of 1980s. Apart from the associated gas that is produced along with crude oil and production of which has been increasing along with increase in oil production, discovery of non-associated gas and offshore and onshore basins has led to growth in reserves and production of natural gas that is more than proportionate to growth in respect of crude oil. Thus from a production figure of 2.4 billion cubic metre in 1981 it has shot up to over 18 billion cubic metres in 1991-92. Similarly the balance recoverable reserves of natural gas went up to about 740 billion cubic metres in 1991-92 from the figure of about around 400 billion cubic metres in 1980-81. A noticeable aspect in respect of natural gas has been the flaring of associated gas. This has been largely due to inadequate evacuation facilities from the oil fields to the consuming centres and lower than the committed off take by the existing customers like fertilizer plants and power stations. During the 7th Plan period (1985-89), more than 40 per cent of the total production of 60 billion cubic metres of gas was flared. The current level of flaring is around 20 per cent. All flaring is targeted to be eliminated by 1996-97 by improved evacuation facilities and supply management. Crude oil and natural gas production targets for 1996-97 i.e. the terminal year of the VIII Plan have been fixed at 50 million tonnes and 30 billion cubic metres. Natural gas has emerged as an important fuel in total hydrocarbon production (oil and gas put together) and it accounted for more than 40 per cent of the total production in 1992-93.

An important characteristic of petroleum product consumption in India is that it has acted as "swing fuel" i.e. oil consumption has grown when other fuels like coal and electricity are not available. This is true of LPG and kerosene consumption in the domestic sector whose consumption has increased when soft coke/electricity are not available and diesel consumption in electricity generation has grown in the face of inadequate/erratic availability of other fuel base electricity and in freight transportation because of inadequate growth in rail transportation. Besides this, with a change in life style there is a switch to petroleum fuels like LPG and greater use of petrol. Consequently from about 17 per cent in 1953-54 the percentage share of oil and gas by 1990-91 has risen to over 43 per cent (and of electricity from 3.3 to 17.6 per cent) while the share of coal decreased from 79.6 to 39 per cent.

India continues to import large quantities of oil and import dependence in respect of oil which was about 50 per cent of the total requirement in 1991-92 is likely to continue at around the same level even at the end of the Plan period as there is a distinct possibility of oil production reaching a plateau of around 50 million tonnes by the end of the current decade, which could also mean an increasingly higher import dependence on oil imports from the beginning of the next century.

As is well known the environmental implications of oil exploration and production arise from the following:

* At the time of drilling, drill cuttings containing drilling fluid, oil and chemical additives are carried to the surface. These cuttings are to be treated before disposal and they can be sources of pollution on the surface and for both surface and sea water.

* At the time of oil production, crude oil carries along with it formation water from underground to the surface. This water contains oil which is to be separated before the water is discharged. Inadequate treatment will affect the environment.

* Crude oil also contains some particulate matters and these matters along with residual bottom sediments and water settle at crude oil tanks and transportation pipelines and is called sludge. This sludge is environmentally hazardous because it contains residual hydrocarbons and its disposal has to be properly organized.

* Flaring of natural gas also leads to pollution because dot emission of sulphur, nitrogen, carbon dioxide etc.

* In respect of oil refining there are both liquid and gaseous effluents and they are causes for potential environmental hazards.

The Central Pollution Control Board (CPCB) has recently evolved a few guidelines regarding oil exploration and production and gas flaring and minimal National Standards (MINAS) have been laid down for effluent discharges from oil production. It is expected that the standards will be followed by commissioning appropriate effluent treatment plants in the next 2 to 3 years. CPCB has similarly laid down the MINAS for effluents and sulphur emissions of certain refining operations. The refineries are expected to conform to these standards by the end 1993. MINAS for sulphur emissions are at present for certain processes and not for the whole refinery operations though some State Pollution Control Boards have set up overall emissions standards for refineries. No standards have as yet been evolved for refinery particulate matters, smoke, carbon, monoxide, hydrocarbons etc. Similarly, no standards have as yet been set up of sulphur contents of petroleum products like diesel, oil and this is an important cause of pollution in the transport sector. CPCB are evolving MINAS in this regard as well. There would now be need for strict observations of the guidelines and MINAS evolved by CPCB which would also need to quickly evolve standards for particulate matters, smoke, carbon, monoxide, hydrocarbons and total sulphur emissions and to enforce them. The standards for petroleum production also need to be evolved quickly since environmental problems are more from the use of petroleum products than from oil production and refining (the sulphur content in HSD in India is 1 per cent as against 0.05 per cent in USA and 0.2 per cent in Switzerland). Also there is need for evolving MINAS for the whole refinery operations. Effective monitoring by State Pollution Control Boards is of utmost importance for containment of environmental damages since the responsibility of enforcement lies with them.[4]

DEMAND SIDE MEASURES

Developing countries have followed the capacity addition approach to meet the increased demand in electricity for their countries while in recent times at least the developed countries have been successful in using demand side management measures (DSMs) to bring down the level of energy consumption without affecting the level of output in the consuming sector.

DSMs perform two main functions :

*   Reduce the demand for electricity on the consumer's premises through improved efficiency where this is economic; and/or

*   Reduce cost by the transfer of energy consumption from times of high demand to times of lower demand where this is economic.

In addition to the benefits resulting from demand and/or energy savings, DSMs can also yield significant environmental benefits through reduced emissions from power stations. DSMs can include energy efficiency equipment such as high efficiency lighting products, or load shifting options such as direct load control using radio teleswitches. DSMs also include other products designed to lead to the efficient use of energy such as cavity wall and loft insulation.

Not all energy conserving measures are DSMs and similarly, not all DSMs will save energy. In this report, we view DSMs as being those applications where direct utility investment is used in order to improve the economic efficiency of the delivery of the services electricity provides. Hence, some measures (such as load management applications) may lead to an increase in energy use but achieve economic efficiency by making use of less costly off-peak electricity. All energy

conserving measures are not necessarily DSMs because the definition we have employed requires cost effectiveness to be considered. One can save energy in a variety of ways, but not all means will prove to be economic.

The introduction of DSMs by a utility or any other party involves costs, benefits and risks. Before any programme of DSMs is introduced, it needs to be demonstrated that the benefits of the programme will outweigh the costs and that the end use customer will benefit. It is also important to establish which parties should bear the risks and share the rewards of investment in DSMs.

Impact of DSMs on the Electricity Sector

The expected consequences for the electricity industry of implementing DSMs will include :

*   a reduction in fuel burning at power plants;

*   the deferral of the capital and financing costs of new power station construction;

*   a reduction in distribution losses;

*   the possible deferral of distribution reinforcement;

*   a reduction in transmission losses;

*   the possible deferral of transmission reinforcement associated with both new power plants and increased loads; and

*   a reduction in the emissions of SO_2 and NO_x both of which are restricted by National Plan limits. In as much as a DSM programme would reduce SO_2 emissions there is an implication in terms of deferring the costs of FGD projects.

Reduced generation costs due to DSMs should be reflected in the electricity pool price, and the price of electricity contracts.

Cost reductions associated with transmission and distribution to DSMs will initially accrue to the firms involved. In the long term, there might be scope for price reduction to the customer to be reflected in changes in the efficiency term of the price control formula. As the costs of maintaining and operating transmission and distribution networks is largely independent of the volume of electricity handled, the cost savings will be small compared with the loss of revenue to these companies. Unit prices for transmission and distribution may have to rise in the longer term if the allowed return to transmission and distribution licensees is to be maintained due to a reduction in the volume of electricity transmitted and distributed.

ENERGY EFFICIENCY

While there is significant potential for energy conservation in all sectors of the Indian economy, the potential for saving is much more apparent in energy intensive industries like iron & steel, fertilizer, textile, cement, pulp & paper, aluminium, ferrous foundry, glass & ceramics and chlor-alkali and in the agricultural sector where over 10 million electrically operated pumpsets have been installed.

The Industrial sector in India has a very high specific energy consumption level as compared to other developing and developed countries. With a significant energy conservation potential of the order of 25 - 30 per cent it offers itself as a prime sector for initiating end-use energy efficiency. The identified energy savings can be achieved through better house keeping, improved capacity utilisation, development of co-generation facilities, industrial waste management etc.

A review of the energy conservation activities at the end of the VII Plan have identified barriers to energy efficiency improvement and a number of activities were initiated to remove the constraints. Some of these are :

i.   Government have taken up a number of schemes to promote energy conservation in the industrial sector, which includes energy audits, building up awareness campaign, training of personnel, conduct of research studies, infrastructural development and setting up of demonstration projects. National Energy Conservation Awards were instituted in 1989 to give recognition to the industrial units for the good work done in improving end-use efficiency.

ii. There is tremendous scope for saving of energy in industrial units in respect to motor drives, their relays and boilers of various sizes. Special attention is, therefore, being given to the evolving of standards, their popularisation and increased adoption by industrial units.

iii. Workshops and seminars are held to discuss issues relating to these as well as arrangements for training of staff involved in the operation of Energy Conservation equipments. Industry-specific programmes, both for managers and workers have also been organised. Expertise from developed countries has been used in these workshops so that the Indian industry can benefit from the experience gained in Western and developed countries.

iv. From a comparison of Indian standards of specific energy consumption figures of dry processing of cement it is seen that the weighted average thermal energy consumption in 1988-89 which was around 867 kcal/clinker has come down to 847 kcal/clinker in 1991-92. Similarly the weighted average for thermal energy consumption in the year 1988-89 for wet process plants which was 1461 kcal/clinker was reduced to 1408 kcal/clinker in 1991-92. A similar reduction in specific electrical energy consumption is also observed, although much lower specific energy indices are reported in some of the progressive Indian cement manufacturing units.

v.   Commissioning of demonstration projects to promote equipments/ technologies.

vi. Providing R & D grant-in-aid for the development of efficient process/equipments.

Besides these, several policy initiatives have been taken by the Government to further uptake the rational use of energy activities in various sectors of the economy.

—   Allowing private participation in power generation this is expected to provide the much needed boost to co-generation.

—   Industries under administrative price control such as cement, paper, aluminium have been decontrolled, consequently inertia which was there for cost reducing measures (due to cost plus pricing) has been overcome leading to major investments being made in these industries for energy conservation: cement is a case in point where large investments have been made to convert energy inefficient wet process plants to energy efficient dry/semi-dry process plants.

—   Specifying minimum economic size of unit for various industries to reap the benefits of economies of scale.

– Disclosure of information on energy conservation in the Annual Reports of 21 industries, under the (Companies Disclosures of Particulars in the Report of the Board of Directors) Rules 1988.

– A decision was taken by the Government in respect of industries under administrative price control that for the period 1982 to 1985 any improvement which results in conservation of energy or reduced consumption of utilities/raw material etc. is made neither the cost nor the benefit thereof would be recognised while computing retention price for a period of six years from commissioning.

– A Standing Committee comprising the representatives of Department of Power, EMC, DGTD, NPC and CII is to study and prepare a list of energy saving devices every year which would recommend to the Ministry of Finance for various fiscal incentives, such as concessional excise and customs duties, 100 per cent depreciation allowance for processing the matter with Ministry of Finance.

Notwithstanding these initiatives, a Cell in the Ministry of Power with the assistance of Energy Management Centre is continuously monitoring the efficacy of existing policies and undertaking micro and macro level studies to formulate and design energy conservation programmes and creating conditions conducive to energy conservation.

Realising that the major constraint to the success of energy conservation efforts so far has been the lack of adequate resources available with users to invest in energy efficient equipment efforts are on to attract both domestic and foreign funds for this purpose. Technical assistance capabilities are also sought to be developed at the State level through setting up of energy conservation cells in the SEBs and State Energy Development Agencies. Efforts of various organisations carrying out research and development activities on energy conservation also would need to be coordinated and strengthened. Legislative support wherever needed to promote conservation is also sought to be given. Development of suitable institutions including strengthening existing agencies at the National, State, District and grass root level is another aspect of achieving higher energy conservation. A national energy usage database is also sought to be developed.

New & Renewable Sources of Energy

It is also being realised that the programme of new and renewable sources of energy is important. The overall scarcity of fossil fuels in India is highlighting this need and the fact that India is a vast country and there are high costs of transportation of energy has made development of locally available renewable and decentralised energy sources to meet growing local needs an important alternative. In the VIII Five Year Plan it is envisaged to develop at least 750 to 1,000 MW of power capacity based on new and renewable sources of energy of wind energy, micro-hydel, urban/agricultural wastes, solar photovoltaic and through cogeneration programmes in process industries such as sugar, paper, food processing mills using biomass and crop residue.[5]

While it is recognised that large utility operated coal based thermal power stations and diesel generators are perhaps more economical today, it can be expected that costs of renewable energy based systems which are at present mostly in the R&D stage or experimental field testing and prototype development stages, will decrease in the near future. The cost of power generation from small hydro stations are already comparable to that from conventional systems. Wind form and bio mass based technologies particularly dendro-thermal and wood gasifier systems are also not far behind. Noting this, the Government is encouraging these technologies by providing financial support for R&D, demonstration and dissemination.

REFERENCES :

1. The Implications of Agenda 21 – an overview, Prodipto Ghosh and Ajai Malhotra, Tata Energy Research Institute (TERI), New Delhi.

2. Coal Sector : Development and Environmental Issues, Mr. R. K. Sachdev, Ministry of Coal.

3. Hydro Power : Development and Environmental Issues, Dr. B. S. K. Naidu, National Hydro Power Corporation, New Delhi.

4. Petroleum Sector : Development and Environmental Issues, Mr. Prabir Sengupta, TERI, New Delhi.

5. Techno-economics of electric power generation through renewable sources of energy : a comparative study, V. V. N. Kishore and K. Thukral, TERI, New Delhi.

CHAPTER IV

ENERGY DEVELOPMENT AND MANAGEMENT IN INDONESIA[*]

Introduction

Energy as one of natural resources is inseparable element to the life of human being. It is critical to economic and social development and without the services it provides, development is not possible both in industrialized and in the developing countries. The prosperity of a nation is often linked to its capability to utilize energy and its level of energy consumption. The utilization of most resources always brings about the negative impact to the environment which causes harm to the life of human being. Utilization of energy, especially those of non renewable fossil fuels, fall into this category. Almost each step of fossil-fuel utilization creates damaging impact to the environment, started from time of fuel extraction to fuel burning process. Ironically, the energy utilization in any country in this world has long been predominated by this non renewable fossil fuels and this utilization pattern is likely to continue at least for several more decades.

Today's world growing environment concerns has lead the way to the efforts of finding more environmentally benign energy utilization technologies as well as energy alternatives for the future. Nearly all countries in the world have included environmental considerations into their energy strategies and development and have tried to find alternative ways to mitigate the energy related environment impacts at the global and national levels.

In relation to the utilization of energy resources, Indonesia definitely subscribes to the strategy of sustainable energy development and management. All possible options are explored including energy conservation, shifting away from the use of fossil fuels as well as recycling of waste heat or of material in industrial processes. National as well as sectoral energy strategies that ensure the minimal demage to the environment while still sustaining the pace of economic and social development have been established. As a developing country, the country realizes that increased use of energy is a must. However, it realizes that the increased energy use without which development is not possible will add enormously to environment destruction.

Therefore the country put the strong efforts that the utilization of energy should be developed wisely; should be for the utmost benefit and welfare of the people, considering also the needs of future generations and most importantly it shall not destroy the environment.

GENERAL POLICY ON ENERGY

Indonesia's energy development during the second long-term (1994 to 2014) development strategy is directed toward achieving self-sufficiency in energy which will be achieved through

a)   developing and maintaining resources of energy

b)   diversifying the use of energy sources

c)   conserving energy use and

d)   developing more intensively the use of renewable sources of energy.

All of the above objectives are in line with the policy of sustainable energy development and the goal of environment preservation. Energy development during REPELITA VI (the Sixth Five-Year Development Plan; 1994/95 – 1998/99) is more specifically directed toward enhancing the welfare of the people as well as meeting the energy needs of the society at large, by guaranteeing the availability of energy and the improvement of quality of service to consumers. Moreover, the energy development should guarantee the long-term sustainability of energy supply, the fulfilment of domestic energy needs, export opportunities, public security and safety, and most importantly the conservation and preservation of environment.

In order for the country to achieve the above long-term policy objectives, a National Policy on Energy has to be implemented. The National Policy on Energy consists of three policy measures as follows:

[*]   Presented by Emy Perdanahari, Head, Sub-Directorate for Energy Programme, Directorate General of Electricity and Energy Development, Ministry of Mines and Energy, Indonesia at the Expert Group Meeting Preparatory to the First Session of the Committee on Environment and Sustainable Development, Bangkok, 30 September – 2 October 1993.

Intensification

Intensifying survey and exploration activities to identify the potential of economically exploitable energy resources (to develop and maintain reserves of energy).

Diversification

Diversifying the use of energy for domestic consumption, aimed at decreasing the role of oil and increasing the development and utilization of other potential, non-oil resources, renewable sources of energy included, with due consideration to the economic, social and technological aspects.

Conservation

Using energy more efficiently and rationally, aimed at safeguarding the sustainability of energy supply and achieving a more balanced development in the pursuit of equity in economic growth and at protecting the environment.

OVERVIEW OF ENERGY SITUATION

As an oil exporting country, Indonesia still largely depends on oil as one of its domestic energy supply (65 per cent in 1992/1993) despite the rapid depletion of oil reserve due to increasing domestic consumption as well as oil's significant role in foreign exchange earnings (30 per cent).

Diversification in the energy sector has not gained significant achievement despite the government's vigorous efforts to develop the use of renewable as well as non-oil renewable resource. Alternative energy resources have not been utilized commensurate with its availability and potentiality. The imbalance in the energy use pattern has probably been caused by firstly the still relatively low price of conventional energy resources, particularly oil which reduces the competitiveness of non-oil resources; secondly the high capital intensity of the installations needed for the conversion of non-oil resources to the required forms, and thirdly, the development of technologies and equipment. To some degree, diversification policy, however, has gained support, especially from the industrial sector which has utilized natural gas and coal and the household sector which has utilized Liquid Petroleum Gas (LPG) to replace the use of kerosene and fuel wood. Presently, non-conventional renewable sources of energy, among them solar, geothermal, wind energy and biomass, are used on a relatively small scale and prevails only in remote areas.

Electric sector has gained substantial achievement in the implementation of energy diversification policy. This was shown by the increasing share of coal, natural gas, hydro and geothermal energy utilized in this sector. There is still more to be done, however, to encourage the use of renewable energy such as hydropower and geothermal in this sector.

In line with rapid economic development and the changing life-style which tends to be more consumptive, the rate of energy demand, mainly for oil, is increasing rapidly. Indonesia's main concern today is the fact that energy is still being consumed inefficiently.

In 1990, energy intensity of Indonesia was 366 toe/million US$ (1980 constant $), while those of other ASEAN countries such as Brunei Darussalam, Malaysia, Philippines and Singapore (in the same year) were 120, 341, 250, and 258 toe/million US$, respectively. The average energy intensity of ASEAN was 285 toe/million US$ (1990's figure, 1980 constant $). During 1985 to 1990, Indonesia's energy intensity has the tendency to increase (see table 1). This indicates that the energy consumption of the country is still inefficient.

Table 1. Energy Intensity in Asean Countries
(toe/million US$ (1980 constant $))

Country	1985	1986	1987	1988	1989	1990
Brunei Darussalam	111	109	118	123	119	120
Indonesia	304	309	310	313	343	366
Malaysia	278	295	299	291	321	341
Philippines	232	221	211	221	235	251
Singapore	170	192	205	224	255	258
Thailand	283	290	290	309	348	374
Asean *)	230	236	239	247	270	285

*) Average Energy Intensity ASEAN

Source: ASEAN Energy Review, AEEMTRC

Based on surveys conducted in several sectors, energy saving opportunities are substantial (see table 2). Energy saving opportunities in energy intensive industries; small-scale and less intensive industries; commercial buildings and household sector were 10-30 per cent, 5-40 per cent; 20 per cent and 20-30 per cent, respectively.

Table 2. Energy Saving Potentials

Sector	Number Surveyed	Energy Saving Potential
A. Industry		10-30%
1. Conducted in Cooperation with		
– Trans Energy – France	44	
– UNIDO	6	
– AEEMTRC	8	
2. Conducted by Ministry of Mines and Energy in home industries, small scale & less intensive industries	130	5-40%
B. Commercial Building		
* Conducted by Ministry of Mines and energy in cooperation with ITB	30	20 %
C. Household	2700*)	20-30%
* Conducted by Ministry of Mines and Energy in cooperation with ESMAP World Bank		

*) Numbers of Household Surveyed
Source: Ministry of Mines and Energy,
Directorate General of Electricity and Energy Development

At the beginning of the First Five Year Development Plan (PELITA I 1969/1970), total commercial energy consumption was 50,064 thousand boe. In line with the escalating growth of development, commercial energy consumption grew substantially and by the end of PELITA V has reached the figure of 449, 102 thousand boe. During PELITA I (1969-1974) and PELITA II (1974-1979), the annual commercial energy consumption grew at 11.6 per cent and 15.3 per cent, respectively. Whereas during PELITA III (1979-1984) and PELITA IV (1984-1989), the annual consumption growth decreased substantially to only 6.8 per cnet and 6.1 per cent respectively. And during PELITA V (1989-1994), the consumption growth has reached 7.7 per cent annually (See Tables 3 and 4).

Electricity Sector

The increasing growth rate of transmission and distribution facilities, electricity sold, electricity production, number of customers and villages electrified were observed during the first long term development plan (PJPT I). During this period, the development in electricity sector has significantly increased. Before PJPT I (1968/69), the total installed capacity was only 536.2 MW. Twenty five years later (1993/1994), the installed capacity has increased to 12,955 MW. During the same period, electricity production has increased from 1,382 GWh to 49,207 GWh. Whereas, the electricity sold by PLN increased from only 1,205 GWh to 41,675 GWh. The development of transmission and distribution facilities were also substantial during this period, the transmission lines facility has increased from only 2,800 km to 19,896 km. Meanwhile, the distribution facilities of Medium Voltage increased from only 5,060 km to 118,315 km and of low voltage increased from 13,400 km to 162,442 km. Facilities of substation has also increased from 1,300 MVA to 23,936 MVA; while the distribution stations increased from 2,300 MVA to 17,889 MVA. The number of villages electrified were also increased from only 2,244 villages in 1978/79 to 29,892 villages at the end of PJPT I (1993).

ENERGY PLANNING AND STRATEGY OF SUSTAINABLE DEVELOPMENT
(The Second Long Term Development Plan, 1994/95-2019/2020)

Energy Sector

Development of energy utilization has so far been directed toward managing energy utilization more rationally and efficiently, commensurate with this increasing energy needs to meet the growing national development. The growth rate of energy consumption however, is expected to grow faster in the future, so that coordinated and overall policy and strategy on energy utilization needs to be established to guarantee the fulfilment of energy needs with the quality and the price that is within the reach of society and most importantly create the least negative impacts to the environment.

During the Second Long-Term Development Plan, energy development is directed in the pursuit of achieving self-sufficiency in energy, and for that purpose the strategy will be focused on preserving energy resources, diversifying energy utilization, rationalizing its use and increasing the utilization of renewable energy sources. Nearly every step of energy activities started from the time of fuel extraction to fuel burning process has the potential for environment damage in the form of water, land or air contamination. Therefore, energy development should always be in line with the strategy of environment preservation.

Energy development is aimed at fulfiling the increasing energy supply need either for the improvement of the quality of life as well as for economic and national development. Potential energy resources, conventional as well as non-conventional, are utilized and developed taking into consideration the economic, technical, social and environmental aspects. To achieve the goal of resource sustainability, development of energy conservation, alternative energy utilization and clean energy technology shall be given priority.

Table 3. Primary Commercial Energy Consumption
PELITA I – PELITA V
(Thousand boe)

Year	Oil	%	Natural Gas	%	Coal	%	Hydro	%	Geo–thermal	%	Total
Pelita I											
1969/70	43,922.9	87.73	3,076.0	6.14	667.1	1.33	2,397.6	4.79	–	–	50,063.6
1970/71	46,933.6	87.74	3,241.1	6.06	733.2	1.37	2,584.7	4.83	–	–	53,492.6
1971/72	50,908.8	87.73	3,481.7	6.00	839.5	1.45	2,801.4	4.83	–	–	58,031.4
1972/73	59,289.6	89.99	3,101.5	4.71	776.3	1.18	2,718.6	4.13	–	–	65,886.0
1973/74	70,564.7	90.91	3,232.0	4.16	599.7	0.77	3,226.2	4.16	–	–	77,622.6
Total	**271,619.6**	**89.03**	**16,132.3**	**5.29**	**3,615.8**	**1.19**	**13,728.5**	**4.50**	**–**	**–**	**305,096.2**
Pelita II											
1974/75	79,533.2	90.46	3,998.0	4.55	662.1	0.75	3,726.3	4.24	–	–	87,919.6
1975/76	88,415.3	89.45	5,878.9	5.95	668.2	0.68	3,881.0	3.93	–	–	98,843.4
1976/77	97,748.5	89.12	7,554.6	6.89	696.8	0.64	3,679.9	3.36	–	–	109,679.8
1977/78	112,219.9	85.93	13,928.4	10.67	794.9	0.61	3,647.0	2.79	–	–	130,590.2
1978/79	129,884.7	83.69	20,741.9	13.36	748.7	0.48	3,827.6	2.47	–	–	155,202.9
Total	**507,801.6**	**87.22**	**52,101.8**	**8.95**	**3,570.7**	**0.61**	**18,761.8**	**3.22**	**–**	**–**	**582,235.9**
Pelita III											
1979/80	140,139.5	80.48	28,454.1	16.34	782.2	0.45	4,743.0	2.72	–	–	174,118.8
1980/81	150,897.1	78.19	34,859.8	18.06	961.2	0.50	6,259.0	3.24	–	–	192,977.1
1981/82	164,954.9	79.28	35,345.4	16.99	1,019.8	0.49	6,759.0	3.25	–	–	208,079.1
1982/83	165,832.6	79.47	35,007.0	16.78	1,120.7	0.54	6,580.0	3.15	143.0	0.07	208,683.3
1983/84	170,298.5	75.32	42,679.7	18.88	1,061.8	0.47	11,636.0	5.15	418.0	0.18	226,094.0
Total	**792,122.6**	**78.43**	**176,346.0**	**17.46**	**4,945.7**	**0.49**	**35,977.0**	**3.56**	**561.0**	**0.06**	**1,009,952.3**
Pelita IV											
1984/85	170,103.1	71.35	52,071.5	21.84	1,767.2	0.74	14,017.0	5.88	435.1	0.18	238,393.9
1985/86	168,105.5	67.33	57,076.9	22.86	6,372.8	2.55	17,661.3	7.07	448.1	0.18	249,664.6
1986/87	169,325.0	63.87	61,629.6	23.25	12,073.0	4.55	21,601.4	8.15	465.9	0.18	265,094.9
1987/88	180,537.2	63.71	64,324.2	22.70	15,720.4	5.55	21,374.2	7.54	1,436.7	0.51	283,392.7
1988/89	190,122.7	62.93	69,864.7	23.12	19,892.0	6.58	20,232.3	6.70	2,018.2	0.67	302,129.9
Total	**878,193.5**	**65.60**	**304,966.9**	**22.78**	**55,825.4**	**4.17**	**94,886.2**	**7.09**	**4,804.0**	**0.36**	**1,338,676.0**
Pelita V											
1989/90	202,863.1	61.94	73,013.3	22.29	25,790.6	7.87	23,823.9	7.27	2,010.5	0.61	327,501.4
1990/91	229,898.6	64.32	76,961.8	21.53	27,356.6	7.65	20,960.0	5.86	2,246.6	0.63	357,423.6
1991/92	245,191.0	64.23	80,348.0	21.05	31,361.0	8.22	22,749.0	5.96	2,094.5	0.55	381,743.5
1992/93	263,098.0	64.81	82,316.0	20.28	32,365.0	7.97	26,177.0	6.45	2,020.0	0.50	405,976.0
1993/94[*]	286,130.0	63.71	94,839.0	21.12	36,645.0	8.16	27,878.0	6.21	3,610.0	0.80	449,102.0
Total	**1,227,180.7**	**63.86**	**407,478.1**	**21.20**	**153,518.2**	**7.99**	**121,587.9**	**6.33**	**11,981.6**	**0.62**	**1,921,746.5**

[*] Estimated figure, the end of March 1994

Table 4. Final Commercial Energy Consumption
PELITA I – PELITA V
(Thousand boe)

Year	Industry	%	Household	%	Transportation	%	Total*)
Pelita I							
1969/1970	8,178.1	21.33	18,059.4	47.11	12,098.0	31.56	38,335.5
1970/1971	9,496.6	23.12	18,450.3	44.91	13,135.1	31.97	41,082.0
1971/1972	11,401.9	24.81	20,528.0	44.68	14,018.6	30.51	45,948.5
1972/1973	14,474.3	27.55	22,238.1	42.33	15,824.0	30.12	52,536.4
1973/1974	13,932.0	23.08	25,004.2	41.42	21,426.1	35.50	60,362.3
Total	**57,482.9**	**24.13**	**104,280.0**	**43.77**	**76,501.8**	**32.11**	**238,264.7**
Pelita II							
1974/1975	17,332.8	24.78	28,869.2	41.28	23,732.3	33.94	69,934.3
1975/1976	20,034.6	25.29	32,866.0	41.49	26,308.7	33.21	79,209.3
1976/1977	24,196.3	27.38	35,389.3	40.05	28,783.6	32.57	88,369.2
1977/1978	30,336.3	29.58	39,411.7	38.42	32,824.8	32.00	102,572.8
1978/1979	36,677.8	30.98	45,277.2	38.25	36,424.8	30.77	118,379.8
Total	**128,577.8**	**28.05**	**181,813.4**	**39.66**	**148,074.2**	**32.30**	**458,465.4**
Pelita III							
1979/1980	42,144.1	31.90	48,560.4	36.76	41,411.7	31.34	132,116.2
1980/1981	48,309.6	32.89	52,835.6	35.97	45,752.9	31.15	146,898.1
1981/1982	50,206.0	31.62	56,777.9	35.76	51,794.7	32.62	158,778.6
1982/1983	50,816.6	32.10	54,551.8	34.46	52,928.5	33.44	158,296.9
1983/1984	58,181.8	35.94	52,195.8	32.24	51,528.8	31.83	161,906.4
Total	**249,685.1**	**32.94**	**264,921.5**	**34.95**	**243,416.6**	**32.11**	**757,996.2**
Pelita IV							
1984/1985	59,842.3	36.55	50,284.3	30.71	53,589.5	32.73	163,716.1
1985/1986	59,701.3	36.65	49,141.0	30,17	54,050.5	33.18	162,892.8
1986/1987	64,039.1	37.15	49,975.4	28.99	58,382.3	33.87	172,396.8
1987/1988	66,696.6	36.65	51,100.9	28.08	64,198.2	35.27	181,995.7
1988/1989	71,360.5	36.75	53,338.2	27.47	69,454.4	35.77	194,153.1
Total	**321,639.8**	**36.75**	**253,839.8**	**29.01**	**299,674.9**	**34.24**	**875,154.5**
Pelita V							
1989/1990	81,821.4	38.02	56,716.8	26.35	76,669.0	35.63	215,207.2
1990/1991	87,944.7	37.61	60,212.7	25.75	85,705.5	36.65	233,862.9
1991/1992	93,278.2	37.45	63,386.2	25.45	92,386.0	37.10	249,050.4
1992/1993	99,689.0	37.77	66,956.0	25.37	97,276.0	36.86	263,921.0
1993/1994*)	108,062.9	38.01	70,765.8	24.89	105,486.1	37.10	284,314.8
Total	**470,796.2**	**37.77**	**318,037.5**	**25.52**	**457,522.6**	**36.71**	**1,246,356.3**

*) Estimated figure, the end of March 1994

Table 5. Installed Capacity by Type of Electric Generation Plants
(MW)

Year	Hydro	Steam Oil	Diesel	Gas Turbine	Geothermal	Total
1968/1969	184.8	108.8	200.6	42.0	–	536.2
Pelita I						
1969/1970	184.8	113.0	201.7	42.0	–	541.5
1970/1971	189.3	100.8	194.3	42.0	–	526.4
1971/1972	186.9	125.0	203.3	42.0	–	557.2
1972/1973	183.9	225.0	213.0	42.0	–	663.9
1973/1974	278.7	225.0	230.3	42.0	–	776.0
Pelita II						
1974/1975	278.7	250.0	266.9	126.0	–	921.6
1975/1976	320.5	250.0	273.9	284.8	–	1,129.2
1976/1977	320.8	250.0	323.0	482.7	–	1,376.5
1977/1978	322.4	250.0	461.5	828.8	–	1,862.7
1978/1979	351.0	556.0	499.4	882.0	–	2,288.4
Pelita III						
1979/1980	378.0	756.2	506.0	896.0	–	2,536.2
1980/1981	378.6	756.2	523.8	896.2	–	2,554.8
1981/1982	398.2	1,156.3	580.8	897.2	–	3,032.5
1982/1983	437.0	1,356.3	664.2	918.5	30.0	3,406.0
1983/1984	536.4	1,556.3	784.3	1,027.9	30.0	3,934.9
Pelita IV						
1984/1985	536.4	2,086.7	859.7	1,096.8	30.0	4,609.6
1985/1986	1,065.2	2,487.0	936.0	1,116.7	30.0	5,634.9
1986/1987	1,240.3	2,487.0	1,326.2	1,116.7	30.0	6,200.2
1987/1988	1,512.1	2,817.0	1,651.9	1,116.7	140.0	7,237.7
1988/1989	1,969.6	3,417.0	1,769.7	1,233.7	140.0	8,530.0
Pelita V						
1989/1990	1,973.0	3,940.6	1,794.9	1,233.7	140.0	9,082.2
1990/1991	2,095.2	3,940.6	1,869.6	1,233.2	140.0	9,278.6
1991/1992	2,115.2	3,940.6	1,946.0	1,213.8	140.0	9,355.6
1992/1993*	2,178.2	3,940.6	2,059.6	2,534.8	140.0	10,853.2
1993/1994**	2,215.0	4,340.6	2,143.0	4,229.6	250.0	13,178.2
Annual average growth rate (%)						
68/69–73/74	8.6	15.6	2.8	0.0	0.0	7.7
73/74–78/79	4.7	19.8	16.7	83.8	0.0	24.1
78/79–83/84	8.9	22.9	9.4	3.1	0.0	11.5
83/84–88/89	29.7	17.0	17.7	3.7	36.1	16.7
88/89–93/94	2.4	4.9	3.9	27.9	12.3	9.1

*) Estimated
**) Projected

Table 6. Electricity Production by Type of Primary Energy
GWh

Year	Hydro Power	Coal	Natural Gas	Geo-thermal	Oil	Total
1968/1969	757.4	–	–	–	624.4	1,381.8
Pelita I						
1969/1970	725.6	–	–	–	686.4	1,412.0
1970/1971	739.4	–	–	–	836.6	1,576.0
1971/1972	776.6	–	–	–	929.6	1,706.2
1972/1973	686.4	–	–	–	1,226.2	1,912.6
1973/1974	903.5	–	–	–	1,385.0	2,288.5
Pelita II						
1974/1975	1,093.7	–	–	–	1,537.2	2,630.9
1975/1976	1,192.4	–	–	–	1,796.7	2,989.1
1976/1977	1,092.2	–	–	–	2,355.5	3,447.7
1977/1978	1,094.7	–	–	–	2,945.9	4,040.6
1978/1979	1,384.3	–	–	–	3,525.3	4,909.6
Pelita III						
1979/1980	1,433.9	–	40.0	–	4,762.8	6,236.7
1980/1981	1,345.4	–	–	–	6,156.5	7,501.9
1981/1982	1,628.5	–	–	–	6,977.7	8,606.2
1982/1983	1,317.8	–	–	77.7	9,253.7	10,649.2
1983/1984	1,816.3	–	73.0	209.3	10,012.2	12,110.8
Pelita IV						
1984/1985	2,117.7	565.3	156.4	217.0	10,565.5	13,621.9
1985/1986	2,989.7	2,079.7	273.9	223.6	10,270.9	15,837.8
1986/1987	4,935.2	3,477.4	282.7	232.3	9,274.5	18,202.1
1987/1988	4,457.4	4,920.3	292.8	719.4	11,169.4	21,559.3
1988/1989	5,226.9	6,304.9	661.2	1,011.9	11,735.2	24,940.1
Pelita V						
1989/1990	6,629.7	9,323.7	804.1	1,006.9	10,969.8	28,734.2
1990/1991	5,674.9	10,634.7	969.4	1,125.4	15,607.3	34,011.7
1991/1992	6,601.0	11,625.9	1,050.6	1,049.5	17,566.7	37,893.7
1992/1993*	8,787.7	10,774.8	1,075.0	1,083.7	19,178.6	40,899.8
1993/1994**	6,709.8	11,680.0	6,839.3	1,205.4	22,772.2	49,206.7
Annual average growth rate %						
68/69–73/74	3.6	0.0	0.0	0.0	17.3	10.6
73/74–78/79	8.9	0.0	0.0	0.0	20.5	16.5
78/79–83/84	5.6	0.0	0.0	0.0	23.2	19.8
83/84–88/89	23.5	82.7	55.4	37.0	3.2	15.5
88/89–93/94	5.1	13.1	59.6	3.6	14.2	14.6

*) Estimated

**) Projected

Energy has to be developed and managed in the pursuit of long term sustainability of energy resources, meeting domestic energy need, fullfilment of export need, national security, and preservation of environment. Energy development is carried out through surveys, exploration, exploitation and utilization of renewable energy as well as efficiency of energy utilization, energy extraction and conversion. To achieve these objectives, an utmost utilization of conventional and non-conventional energy resources combined with development of applicable energy efficient equipment and technology is a must. And this should be included in the whole frame work of national energy policy.

New and Renewable Sources of Energy (NRSE) such as, geothermal, hydro, biomass, solar, wind energy needs to be intensified taking into consideration the economic, technical feasibility, applicability and detrimental impacts of its utilization. Rural energy development needs to be increased especially through promoting utilization of locally available energy resources combined with efforts to increase the participation and contribution of the local people.

The growth of energy consumption needs to be decreased by way of conservation. This should be applied to both oil and non-oil energy sources. Appropriate regulations and measures combined with application of more efficient energy appliances and technology as well as energy management training for energy users will be of great importance to reduce the consumption growth.

ELECTRIC POWER SUB-SECTOR

The development in the electric power sector is directed toward accelerating economic activities to increase the welfare of people in urban and rural areas. The installation of electric power generation facility is done by government, private sectors and cooperatives. Power sector management should be done in efficient way and should guarantee continuous and adequate supply of electricity; provide a good quality of services, with the price within the reach of society or at fair price to consumers as well as to producers but still guarantee the sustainability and the security of electricity supply and distribution system. Efficient utilization of energy sources for electricity generation is a must and a vital element in the planning of electric power sector in the country.

Electricity generation plan is based on the least cost criteria which subject to, among others, less oil use and more application of environmentally sound technology. Based on that policy, during the last two decades the role of oil fired power plants has been decreasing replaced by coal or natural gas fired thermal power plants. The use of coal, of course, is subject to the environmental regulation which is considered more stringent than other countries.

The implementation of rural electrification programme is being intensified to encourage economic activities, improving the welfare of rural people. Rural electrification programme is developed utilizing local energy sources such as microhydro, wind energy, solar energy and biomass. This will help support energy diversification programmme to shift away from oil dependence and simultaneously encourage the use of more environmentally benign energy sources. Rural electrification programme also provide opportunities to society through local cooperatives to participate in the programme.

In line with the objectives of policy of the electric power development, the main elements of the strategy are, inter alia, as follows:

a   Electric power development is a part of National Development Programme. Therefore, its activities have to be in conformity with the national development programme meaning that every stage of electric power development activities should support the achievement of the goal of National Development, either in improving the welfare of the people or in increasing the economic growth.

b.   In the pursuit of fair distribution of the result of national development programme or of the wealth of the country, domestic electricity need should be met, especially the need for electricity in rural areas.

c.   Electric power development is an integral part of national energy policy which consists of a set policy measures designated as intensification, diversification and conservation. Therefore, every stage of the development should be in line with those policy measures.

d.   In the development and installation of electric power generation plant, the utilization of domestic products as well as services should be pursued to the maximum extent possible.

e. The development of electric power sector should support and be commensurate with the policy of environment preservation, regional development and other national development policies as well.

NATIONAL TARGET ON ENERGY DEVELOPMENT DURING THE SIXTH FIVE YEAR DEVELOPMENT PLAN (REPELITA VI) AND THE SECOND-LONG TERM DEVELOPMENT PLAN.

The main aim of energy development during the second Long Term Development Plan is to maintain the balance between supply and demand by adopting the following principles.

a. Energy resources utilization especially those of fossil fuels should be managed in such a way that it would produce higher value added, either as energy or as raw material.

b. Energy resources utilization during the next two decades is expected to have reached more productive and more efficient use due to the more widespread application of new technology, fair distribution of development activities and better allocation of industrial centers close to energy source.

c. Energy resources utilization pattern should be gradually extended from the existing one of shifting away from oil to a pattern of shifting away from non-oil fossil fuel resources.

d. Energy resources development is still mainly directed toward meeting domestic energy need although there are still possibilities to import energy for economy and strategic considerations.

e. Energy resources should be utilized in more environmentally sound manner and preservation of environment should be done not only from its utilization aspect but also from the aspect of its sustainability.

f. The application of more sophisticated technology on renewable energy (or what is called "future technology") will be materialized. The extent of utilization of energy resources whose potential are marginal and supply system are decentralised will be maximized.

g. The socialization of energy efficient manner supported with the environmentally friendly technology and facilities should be intensified.

h. Enhancement of knowledge in science and technology in the field of energy, especially those of sophisticated and advanced energy technology.

i. Invite the private sector participation to invest in the electric power generation, operation and distribution to meet the fast growing demand of electricity in the country.

In the pursuit of decreasing dependence an oil resources, the implementation of energy diversication and conservation should be intensified. Energy diversification target in the power sector has been significantly achieved. In the transportation sector, the use of CNG has been introduced, although the socialization has not been effectively done. Various renewable energy technology, especially those of small scale have reached commercial stage. Several programmes on energy conservation have been implemented. However, the programmes and activities need to intensified and widely implemented considering the relatively big potential of saving especially in industrial sector.

The main target to be achieved at the end of the second Long-Term Development Plan is self-sufficiency in energy including electricity. Following are detailed targets:

a) all oil and gas/hydrocarbon basins should be explored.

b) oil production should be maintained at the rate of around 1 million barrels per day.

c) the achievement of condition where the oil-gas industry is capable of supporting the development of national industry without sacrificing its own economic development.

d) domestic refineries and EXOR should increase by 4 times the capacity current capacity.

e) Natural gas utilization should increase by 1.5 times the current demand.

f) Petrochemical refinery capacity (upstream) should increase by 5 times at the minimum.

g) Geothermal energy should be utilized at least 25 per cent of the current potential reserves.

h) The role of coal in the national energy mix should be increased commensurate with its potential availability.

i) New and Renewable Sources of Energy (NRSE) like biomass, solar, wind energy should quantitatively play more important role in the national energy mix, either in terms of its utilization for generating electricity as an integral part of national transmission and distribution system as well as for direct use (solar heating system etc).

j) Electrification ratio should reach 100 per cent (shared by PLN, cooperatives, private sector and other electric power producers).

k) The growth of energy demand should be lower than that of the Gross Domestic Product (GDP).

l) The growth of the country energy demand should be lower than that of world's demand.

m) Total primary energy demand is projected to reach 4 times the current demand or equal to 1,407 Million BOE at the minimum.

n) Commercial energy demand per capita is projected to reach 3 times the current commercial energy demand per capita (1.9 BOE) or equal to 5.6 BOE.

o) Primary energy demand of electricity generation is projected to increase by 10 times the 1990/91 condition or equal to 711 million BOE or 50 per cent of total energy demand.

p) Implementation of energy conservation programme is expected to reduce the growth of annual energy demand to 3.5 per cent in average.

q) Enhancement of regional cooperation on the utilization of energy resources especially electricity and gas.

ACTION PROGRAMMES DURING REPELITA VI AND THE SECOND LONG-TERM DEVELOPMENT PLAN (PJPT II)

Following are action Programmes for Energy Development during REPELITA VI and PJPT II

a. Reduce the share of oil and increase the share of alternative energy in the domestic energy supply mix. It is planned that at the end of REPELITA VI, 30 per cent of kerosene consumption in household sector be replaced by coal briquettes.

b. National energy conservation programme will be intensively implemented so that at the end of REPELITA VI, national energy consumption could be reduced by 15 per cent of the consumption level consumption in that year (12.75 per cent of saving in household; 14.5 per cent of saving in transportation and 17.75 per cent of saving in industrial sector).

c. Increase renewable energy utilization especially biomass, wind, microhydro and solar energy so that at the end of REPELITA VI, the utilization should be able to reach commercial stage and should have more significant role in the total energy supply mix.

d. Coordinated rural energy development through the establishment of rural energy planning and rural energy supply programme using local energy sources in remote or isolated areas.

Electric power development programme during Repelitas within the Second Long-Term Development Plan is directed toward the fulfilment of national demand of electricity. To achieve this objective, several projects on the construction of electric power generation facilities including transmission and distribution system will be carried out. The construction of these facilities will be done by the government which also invites private sectors and cooperatives to invest on these projects. During REPELITA VI, approximately 9,523 MW capacity of electric power generation plant, 10,658 kms of transmission lines, 133,319 kms medium voltage distribution lines, 30,976 MWA sub stations, and 21,817 MVA distribution sub-stations will be built. Whereas, private sectors will finance the construction of 2,945 MW capacity of electric power generation plants.

Development of rural electrification during REPELITA VI is expected to increase the number of villages electrified by 18,619 villages. Additional electric power generation in the amount of 50 MW is expected to be provided. Accordingly, the number of villages electrified at the end of REPELITA VI will reach 48,500 or approximately 80 per cent of the total number of villages in Indonesia.

Other energy programmes, for example efficiency improvements, or reduction of distribution losses, implementation of Demand Side Management (DSM), improvement of quality and realibility will be carried out more intensively. Human resources development programme through training both practical and theoretical skill or capabilities, research and development will also be enhanced.

Efforts to reduce environmental impacts caused by energy sector activities, especially those caused by electric power sub sector through Analysis of Environmental Impacts (AMDAL), and public training, e.g on the effects of electromagnetic radiation on human health.

The participation of private sector will be encouraged to invest in electric power generation either on the solicited or unsolicited projects. Besides, the private sector will also be given opportunity to contribute in the local content of electricity projects. International cooperation (bilateral as well as multilateral ASEAN, or other multilateral cooperation) will be enhanced through expertise, information exchange etc.

RELATION BETWEEN ENERGY AND ENVIRONMENT

Energy sources can be classified into two major categories, fossil and renewable energy. Fossil energy sources is a finite resource or the rate of its depletion is much faster than the rate of its formation that takes hundred of thousands to millions of years to form. Renewable energy is an energy source that are continually replaced on cyclical basis with no foreseeable termination of supply. The utilization of energy, particularly those of fossil energy bring harmful impacts to the environment. Each step of fossil-fuel utilization, started from the time of fuel extraction to fuel burning process creates damaging impacts to the environment.

Environmental impacts related to oil utilization

Nearly every step of oil activities e.g. exploration, production, transportation, conservation and consumption, create environmental impact. Exessive quantities or crude oil burned during the process of drilling produce million tons of pollutants. The offshore rig activities produce potential contamination to beaches, damage marine wild life and shoreline fauna caused by oil spills. The environmental damage due to the possible collision during the transportation of crude oil or the damaging tankers as a results of hurricanes is also great. Possible leaks of oil during the transportation of oil through pipelines under the sea water cause great damage to the environment. Crude oil conversion to petroleum products in refineries also produce dangerous effluents such as ammonia, phenols, napthalene, cresols, sulfides, and other noxious chemicals. Burning of oil in industries as well as oil generated power plants produce wide variety of pollutants such as total suspended solids, CO_2, SO_x and gasseous hydro carbons carried higher into the atmosphere by smoke stacks and much of these gaseous pollutants remains suspended in the atmosphere to form "smog" that is eventually descended into the land and into water supplies in the form of acid rain.

Environment impacts related to natural gas utilization

Natural gas is considered as the most environmentally sound energy among fossil-fuel energy sources. The process and handling of natural gas such as purification, compression, liquefaction and transportation often raise more safety concern rather than environment concerns. Like other fossil fuel, however, its consumption will also create pollutants such as unburned CH_4, CO and relatively insignificant amount of CO_2 emission due to its relatively smaller percentage of atom C than atom H, and other gaseous hydrocarbons.

Environment impacts related to coal utilization

The utilization of coal creates environmental impacts from the time of extraction to long after it is burned. There are several environmental issues concerning underground mining, ranging from the working environment to the impact of mining activity of the site and adjacent area. Coal miners are also exposed to many health and safety risks, from noxious, explosive, and toxic gases to caveins and block lung disease. Pure water supply or water naturally found in the mine is easily contaminated by sulfur, acids and coal dust or other contaminants. Water after used for extraction process should also be removed, because it contains toxic impurities. Other problems such as sinkholes, lowering water table of the mining area are also created by the excavation activities. The preparation and cleaning process of coal commonly carried out near the mine after extraction also bring about another source of natural water contamination. The preparation such as grading, crushing, sizing and cleaning process require large quantities of water, which can pollute the adjacent aquifer, cause water degradation, erosion etc. The piles of coal stored in adjacent area is the source of water pollution. Burning coal to convert its chemical energy to heat energy (combustion process) causes a far greater concern of environment degradation. The combustion gases such as carbon dioxides, sulfur oxides, NO_x and particulates or fly ash, the unburned materials released through smoke stacks creates environmental degradation. When sulfur in the coal is combined with the oxygen in the air during combustion, SO_2 is formed. SO_2 which is released through the stack combined with water vapor (H_2O) to form sulfuric acid (H_2SO_4), commonly called acid rain. Nitric acid (NO_x) which is formed when heating of air during combustion of coal, together with sulfuric acid, hydrocarbons and heavy metals are the source of forest damage, and causes pollution of watersheds leading to the biological damage of adjacent lakes, rivers etc. The excessive CO_2 emission from coal combustion process may cause a permanent global warming trend as a result of the "greenhouse effect".

Environmental impacts related to hydro-power utilization

The dams and reservoirs built for large hydro-electric facilities may create a number of environmental impact, some positive, some negative. The positive impacts are the building of recreational areas for tourist, fishing, boating which create additional regional income. Another positive aspect is that of the removal of a hazard caused by devastating floods to people living in the adjacent area. Instead, soil erosion can be controlled and water for irrigation can be easily provided. The negative impacts would be the building of reservoirs bring about the loss of natural existing scenic areas, residential areas, fertile farmland, wild-life areas. Over a long period of time, the reservoirs have a tendency to fill with silt sedimentation and become less effective as generating facilities.

Environmental impacts related to geothermal energy utilization

Some scientists are calling geothermal energy the safest, cleanest, and most economical of the energy alternatives being considered for commercialization. A number of issues have been raised concerning the environment impacts on the adjacent area of geothermal facilities. Most of the potential environmental impacts of geothermal energy are relatively minor, short lived and affect only the immediate site of the facility. Thus, geothermal energy utilization is relatively clean, however, some associated problems exist. The

development of a site for electricity generation surely causes some problems, such as the changing of the land use, disrupt the normal residential routines, and affect adjacent agricultural efforts.

Environment impacts related to biomass utilization

In spite of some serious mismanagement and over cutting of woodlots and forests, wood as well as other biomass energy is renewed in less than a generation, and therefore hold promise for our energy future. The environment impacts of solid biomass combustion is somewhat different than that of fossil fuels. Burning biomass carefully controlled burners results in insignificant sulfur and nitrogen oxide emissions compared with those of fossil fuels. CO_2 levels in the atmosphere are essentially the same when wood is burned as when it is allowed to decay on the florest floor or absorbed by plants, although the amounts of CO_2 are produced in a much shorter span of time. Increased use of wood-burning stoves in homes, if not carefully controlled, may cause significant increases in particulate matter released into the atmosphere. An additional concern is that policyclic organic matter may make up to 4 or 5 per cent of organic emissions, some of which have been shown to be carcinogenic.

Environmental impacts related to liquid biomass utilization

Alcohol fuels burn cleanly in existing engines and produce the same power and mileage as gasoline, depending on the individual engine, drive and driving conditions. Emissions are reduced for HC and CO, although NO_x levels are slightly increased. There are some disadvantages to the use of alcohol fuels. Alcohol has a solvent property which may attack some rubber and plastic parts (elastomers). This property may harm some auto finishes and cause various gaskets and seals to decompose. Water free alcohol is expensive to produce, and if alcohol is less than 99 per cent pure, there may be a phase separation (water/alcohol) which could cause combustion problems.

Environmental impacts related to gaseous form of biomass utilization

There is relatively minor environmental problems associated with the operation of gasification units. The gas contains CO, which has no taste or smell and could be breathed with serious or fatal consequences. When the unit is opened for charging it with wood, it could expose the operator to high heat and flare-up Unit cannot be operated in buildings or enclosed spaces within adequate mechanical ventilation.

Environmental impacts related to methane gas utilization

Methane gas is colorless, tasteless, and odorless when there is no H_2S component. Small amounts of the latter can serve as a check on possible leaks of the system. Biogas, with its low carbon dioxide content, is nontoxic, but like many other gases it can cause asphyxiation if allowed to accumulate and replace oxygen in an enclosed space. The utilization of biogas from biomass or animal waste will contribute to the reduction of CH_4 level in the environment. Instead of allowing methane gas produced from farm waste or biomass waste pollutes the environment, it can be utilized as one of renewable resources. The utilization of biogas would also solve another major environment problem caused by the accumulation of organic waste.

Environmental impacts related to solar energy utilization

The utilization solar energy is considered to be one of the safest and cleanest of all the energy options. However, there are several environmental and safety concerns associated with manufacturing of semiconductor, silicon made from silica, one of the most common materials found on earth. The danger of breathing of silica dust by miners is as imminent as that coal dust exposed to the coal miners. The extraction and refining of ores for the dopant materials could produce a large amount of mercury and alumina sludge residues. The production of semiconductor also use toxic gases which often cause danger to the workers. Workers of photovoltaics are exposed to several acids and poisonous gases that have found to cause burns and respiratory illness.

Environmental impacts related to wage energy utilization

The environmental impacts of wave-energy-conversion systems is minimal. Environmental concerns have been associated with the potential interference with fishing and offshore oil exploitation efforts, shipping, the physical state of the adjacent coastline, spawning of fish, and the human visual and social amenities.

Environmental impacts related to OTEC utilization

Environmental impacts associated with OTEC units have proven negligible to date, with no serious problems foreseen in the future. OTEC utilization could create both positive and negative impacts. The negative impacts would be the pollution of ammonia, the toxic working fluid used for closed systems. However, it is easily detectable and biodegradable. The production of titanium-alloy heat exchangers posess some concern because of the mineral limited availability and when react with magnesium or sodium metals, it produces high-temperature reactions, then, causes explosion. Many OTEC plants discharging cold water could lower the ocean's surface temperature, affecting the evaporation and CO_2 exchange rates in the adjacent area. Cool water from the depths of the ocean which is rich in nutrients and living organism (plankton).

Environmental impacts related to wind utilization

The wind-driven systems produce no pollutants and are environmentally safe. Improvements in blade design have reduced noise to the levels no greater than the noise of moderate to high winds.

THE MANAGEMENT OF ENERGY RELATED ENVIRONMENTAL IMPACT AN OUTLOOK FOR THE FUTURE

National energy strategy is directed toward achieving both higher economic development of the nation while taking environmental impacts and prevention cost fully into account. The lack of the available information and uncertainties on the level of impacts, a poor understanding on the abatement technology and other issues have made the action plan on the pollution control AND preventive action difficult to implement.

Following are recommended plan of action for the management of energy related environment impacts.

a. Acceleration on the development of cleaner fuels utilization

Investment in cleaner fuels such as natural gas need to be given priority. The use of CNG in transportation in Jakarta and Surabaya sector should be increased and should be introduced in other major cities as well. A study by LEMIGAS indicates that CNG has the potential of replacing about 76 per cent of gasoline and 99 per cent of automotive diesel used by road transport in Jakarta. Another possibility is to reduce the lead content of gasoline by providing an incentive for the accelerated introduction of unleaded or low lead gasoline.

b. Establishment of a policy framework for environmental responsibility

The implementation of environment management strategies of Indonesia has so far been hampered by the lack of a national enforcement strategy or programme, limited resources in skilled and trained inspectors, limited funding and no-clear definition of responsibilities among central departments in the prevention and management of air pollution. The introduction of disincentives for those who create pollution or charges for the disposal of their emissions is of great importance. The polluter pays principle should be implemented. They could choose whether to pay for the damage they have caused associated with the use as well the producing of fuel or to reduce the pollution which will be made cheaper. This will require tremendous efforts and overall studies to bring it into reality as there is still great uncertainty about the accuracy of the damage estimates.

c. Energy efficiency improvement and pollution control

Pollution control and efficiency improvement in all energy related sectors need to be accelerated. Sectoral proposed measures are listed as follows:

(i) Industrial sector
- Improved efficiency in production processes
- Recycling of waste heat in production processes

(ii) Transport sector
- Improved fuel efficiency of vehicles
- Promotion of public transport

(iii) Residential/Commercial Building Sector
- Improved energy utilising equipment and systems
- Improved lighting
- Co-generation

(iv) Electricity Generation
- Increased efficiency of fossil-fuelled power plants
- Co-generation of electricity and heat.

d. Priorities for Research and Analysis

Improved research and analysis is of great importance for understanding the magnitude of environmental impacts of energy utilization. Indentification of the causes and costs of impacts, assessment of the damages to environment are required to formulate policies and strategies which focus on the areas of greatest need and guarantee the effectiveness of abatement. The following areas need to be further given attention.

- Air Quality Management Strategy
- Additional Epidemiological Studies
- Consideration of Local Conditions in the AMDAL Process
- Petroleum Pollution Monitoring
- Marine Ecosystem Sensitivity Studies
- Prevention and Management of Pollution from Shipping
- Strengthening AMDAL for the Mining Sector
- Assessment of Renewable Energy Options
- Climate Change Mitigation

CONCLUSIONS

From what have been discussed above, the following conclusions can be made:

(1) Energy development in Indonesia is aimed at guaranteeing self sufficiency on energy.

To achieve this goal, efforts should be enhanced

(a) to develop and maintain resources of energy

(b) to diversify the use of energy sources

(c) to conserve energy use

(d) to develop more intensively the use of renewable sources of energy

(2) Energy development should take into account the long-term sustainability of energy supply, the domestic energy needs, export opportunities, public security and safety, and conservation of the environment.

(3) The main strategies for sustainable energy development and management in Indonesia are:

(a) Increasing efficiency of energy utilization in all economic sectors.

(b) Decreasing the role of fuel-oil in national energy mix and increase the use of environmentally sound energy sources such as natural gas, hydro, and geothermal.

(c) Utilizing renewable energy to the maximum extent possible taking into account the economic, social and technological aspects.

(d) Implementing demand side management (DSM) programme in the electricity sub-sector.

(e) Encouraging society especially those living in rural areas to meet their own demand of energy including electricity by utilizing local renewable and more environmentally clean non-oil energy sources.

(f) Encouraging newly established industries to fulfill energy needs by utilizing locally available energy sources and, whenever possible, adjust or select production processes that are most appropriate based on the non-oil more environmentally sound energy sources.

REFERENCES

(1) "Overview of Energy Situation and Strategy of Sustainable Development in Indonesia", A. Arismunandar, Jakarta, September 1993.

(2) "A Survey of the Energy Situation in Indonesia" A. Arismunandar and Widartomo, Jakarta, September 1993.

(3) "Draft of Repelita VI, Energy and Electric Power Sector, Directorate General of Electricity and Energy Development, Ministry of Mines and Energy, Indonesia, Jakarta, September 1993.

(4) "Indonesia's Electricity Sector Development Programmes and Private Participation", Nengah Sudja, Jakarta, September 1993.

(5) "Indonesia, Energy and The Environment: A Plan of Action for Pollution Control, Country Department III, East Asia and Pacific Region, June 1993.

(6) "Energy and Environment: Impacts and Controls", United Nations, Department of Technical Cooperation for Development (UNDTCD), October 1990.

(7) "The Study on the Response Actions Against the Increasing Emission of Carbon Dioxide in Indonesia", Japan Environment Agency, March 1993.

CHAPTER V

KOREAN ENERGY POLICY FOR THE CLIMATE CHANGE CONVENTION[*]

1.   Introduction

Global warming has aroused intense concern in the Republic of Korea since the Rio Climate Change Convention. The Convention ultimately seeks to control and reduce the utilization of fossil fuels, which have been a major source of energy. Since the Republic of Korea, as a newly industrialized nation, has long experienced a rapid increase in energy consumption, with an average growth rate of 8.4 per cent per year for the 1970-1992 period and of 10.9 per cent per year for the 1985-1992 period, the Climate Treaty is expected to have serious impact on the various energy and economic sectors in the Republic of Korea. As a result, it now needs to evaluate seriously the implications of the Convention in order to participate in the concerted effort to address climate change.

The first part of this paper examines the current status of energy demand and CO_2 emissions in the Republic of Korea and presents the results of a "Business-As-Usual" long-term scenario for the year 2030. Then we roughly estimate the amount of potential reduction of GNP growth rate in case that the emission level is set to 1990 EC per capita emission level. Finally we consider the various response strategies for the

Republic of Korea.

2.   Trend and Future Projections of Energy Demand in the Republic of Korea

a.   Trend of Energy Use in the Republic of Korea

The rapid industrialization of the Korean economy resulted in a high increase in energy consumption. Having nearly doubled every 10 years, total primary energy consumption has increased more than 8 times over the last 3 decades, reaching 116.0 million TOE in 1992. Per capita energy consumption in the Republic of Korea increased from 1.2 TOE in 1980 to 2.7 TOE in 1992, which is, however, still far below the average of the developed countries. For purposes of comparison, in 1988, per capita energy consumption was 7.9 TOE in United States, 3.3 TOE in Japan.

Energy intensity, an index of the energy required to produce a unit of value added, has increased since 1989 as energy demand has increased faster than economic growth. The high increase in energy demand resulted mainly from the expansion of energy-intensive industries and an increase in the number of motor vehicles.

Table 1. Major Energy Economic Indicators

	'80	'85	'92	80-85	85-92
GNP(Trillion won,1985)	52.3	78.1	148.3	8.4	9.6
Share of Manufacturing	27.5	31.4	34.1		
Energy Demand					
Primary Energy (Mil.TOE)	43.9	56.3	116.0	5.1	10.9
Per Capita (TOE)	1.2	1.4	2.7	3.1	9.8
Oil (Mil.Barrels)	182.1	189.2	514.2	0.8	15.4
Energy Import Dependence					
(nuclear included) (%)	73.6	76.2	93.6		
(nuclear excluded) (%)	71.6	68.8	81.4		
Import Bill for					
Energy (Bil.US$)	6.5	7.3	14.5	2.3	10.3
Oil (Bil.US$)	6.0	6.1	12.1	0.3	10.3
Energy/GDP (TOE/1000US$)	0.71	0.61	0.68		

[*]   Presented by Kiho Jeong, Research Fellow, Korea Energy Economics Institute Republic of Korea, at the Expert Group Meeting Preparatory to the First Session of the Committee on Environment and Sustainable Development, Bangkok, 30 September - 2 October 1993.

Table 2. Changes in Some Economic Activities in the Republic of Korea

	'80	'85	'90	'92
Productions in Energy-Intensive Industries				
Pig Iron (Mil.M/T)	5.6	8.8	15.3	19.2
Cement (Mil.M/T)	15.6	20.5	33.9	44.4
Ethylen (Thous:M/T)	368.2	560.9	1 064.9	2 811.5
Number of Motor Vehicles (Thous.)	527.7	1 113.4	3 394.8	5 230.9
Passenger Cars	249.1	5 56.7	2 074.9	3 461.1

b. Long-term Outlook of Energy Demand in the Republic of Korea

The issue of global warming, the control of energy use, and economic growth are closely interrelated. An evaluation of the energy demand outlook in the long-term perspective is essential in identifying the potential in addressing climate change. As frequently done in other studies, the long-term energy demand scenario up to the year 2030 is approximated using various assumptions, i.e., that the current trends of economic growth and energy consumption are maintained and published government plans for nuclear power and natural gas are implemented. This scenario might be called a "Business-As-Usual scenario" (BAU).

GNP is expected to grow 6.6 per cent in the period of the 90's and 4.5 per cent during the period from 2000 to 2030. The fuel economic growth, primary energy consumption is expected to grow 6.7 per cent and 2.7 per cent in the respective periods.

Consequently, energy consumption will be 1.9 times and 4.2 times the level of 1990. Coupled with a relatively low population growth rate of less than one per cent, per capita energy consumption is 2.1 TOE in 1990, 3.8 TOE in 2000, and 7.8 TOE in 2030.

c. Projection of CO_2 Emissions in the Republic of Korea

Based on the BAU projection of energy consumption, CO_2 emission is projected to be 1.8 times its level of 1990 in 2000, and to be 3.4 times in 2030. In the year 2030, carbon emissions will be 227 million carbon tons.

With low population growth and continued economic growth, per capita emission of CO_2 will continue to grow such that after 2000, it will surpass the EC average level of 1990, 2.4 carbon ton per capita. The amount of CO_2 per unit of GNP, expressed in ton C/Million Won, is expected to decrease 2.4 per cent annually after 2000.

Table 3. Energy and Economic Projections under BAU Scenario

	1990	2000	2010	2020	2030	
GNP (Thou.Bill Won)	130.4	248.2	423.9	658.4	928.7	
Population (Thou.)	43,520	46,828	49,486	50,193	50,193	
Energy Demand (Mil.TOE)						
Primary Energy	93.2	177.7	253.3	327.4	392.1	
Final Energy	75.0	140.8	193.3	241.1	282.3	
Energy/GNP (TOE/Mil.Won)	0.71	0.72	0.60	0.50	0.42	
Energy/Population (TOE)	2.14	3.80	5.12	6.52	7.81	
Fossil Fuel Share (%)	83.3	85.2	79.2	75.6	73.4	
Energy/Gnp Elasticity		1.0	0.7	0.6	0.5	
average growth rate (%)		90-00	00-10	10-20	20-30	00-30
GNP		6.6	5.5	4.5	3.5	4.5
Population		0.7	0.6	0.1	0.0	0.2
Primary Energy		6.7	3.6	2.6	1.8	2.7

Table 4. Indicators of CO_2 Emissions

	1990	2000	2010	2020	2030	Growth rate (%) 90-00	00-10	10-30	00-30
CO_2 Emissions (Mil.TOE)	67.1	121.8	157.7	194.8	227.1	6.1	2.6	1.8	2.1
Per Capita Emission (Carbon Ton)	1.6	2.6	3.2	3.9	4.5	5.0	2.1	1.7	1.8
CO_2/GNP (Carbon Ton/Mil.Won)	0.46	0.45	0.34	0.27	0.22	−0.2	−2.8	−2.2	−2.4

Table 5. Share of CO_2 Emissions by Fuel Type (%)

	1990	2000	2010	2020	2030	Growth rate			
						90-00	00-10	10-30	00-30
Oil	53.7	54.8	51.6	47.8	44.0	6.3	2.0	1.0	1.4
LNG	2.6	5.8	8.8	10.3	11.5	15.2	6.9	3.2	4.4
Coal	42.6	38.3	37.8	39.0	40.2	5.0	2.5	2.2	2.3
Anthracite	17.4	4.3	1.3	0.5	0.1	−7.6	−9.2	−11.6	−10.8
Bituminous	25.2	34.0	36.5	38.5	40.2	9.3	3.4	2.3	2.7
Renewable	1.1	1.0	1.8	2.9	4.2	5.1	8.6	6.3	7.0

Table 6. Share of CO_2 Emissions by Sectors (%)

	1990	2000	2010	2020	2030	Growth rate			
						90-00	00-10	10-30	00-30
Energy Transform	15.7	24.3	25.9	29.3	32.0	10.9	3.3	2.9	3.0
Electricity	15.7	23.7	25.1	28.2	30.7	10.6	3.2	2.9	3.0
District Heating	0.0	0.5	0.8	1.1	1.3	–	6.4	4.5	5.1
Final Consumption	84.3	75.7	74.1	70.7	68.0	5.0	2.4	0.6	1.7
Industry	36.6	35.5	37.7	39.5	40.6	5.8	3.3	2.2	2.6
Transportation	17.0	22.0	21.4	19.1	17.5	8.9	2.4	0.8	1.3
Resid. & Commerce	27.9	16.8	13.9	11.2	9.1	0.9	0.7	−0.3	0.0
Public	2.8	1.5	1.1	0.9	0.8	−0.4	−0.3	0.1	−0.0

The major source of emissions of CO_2 was coal until 1988. After 1990, oil has become the primary source of carbon emissions, with more than 50 per cent. Emissions of CO_2 from bituminous coal will rapidly increase, with its share rising from 25 per cent in 1990 to 40 per cent in 2030.

Emissions of CO_2 by sector shows that after 2000, the power generation and industry sector will become significant contributors. The annual rate of increase in the 1990's and the 2000's would be 11 per cent and 3 per cent, respectively, for the power generation sector and 6 per cent and 3 per cent, respectively, for the industry sector, resulting in sector's share in 2030 being 32 per cent for power generation sector and 41 per cent for industry sector. The high growth in the power generation sector will be largely due to the increasing use of bituminous coal in power generation. The growth in the industry sector will be due to its large dependence on oil. The CO_2 emission by the transportation sector will increase with a high rate of 9 per cent in the 1990's but with a low rate of 1 per cent in the 2000's. The residential and commercial sector will undergo structural changes in its use of fuel so that the sector's share will drop from 28 per cent in 1990 to 9 per cent in 2030.

3. Required CO_2 Abatement and GNP Growth Rate Reduction

Total CO_2 emission of the Republic of Korea is 67.1 Mil. ton of carbon in 1990 (1.5 carbon ton of per capita emissions). This figure is expected to be expanded by the energy demand for economic growth in BAU scenario. To meet average per capita CO_2 emissions of EC (2.4 carbon ton per capita), it will be needed to reduce 7.7 per cent of total carbon emissions in 2000 (9.4 Mil. Carbon tons) and 24.8 per cent in 2010 (39.2 Mil. carbon tons), compared BAU case.

Table 7. Required CO_2 Abatement and GNP Growth Rate to Meet the 1990 Emissions Level

	2000	2010
CO_2 Emission (BAU) (Mil.TC)	121.8	158.0
Required CO_2 Abatement (Mil.TC)	54.7 (44.9%)	90.9 (57.5%)
GNP Growth rate Reduction (%)	3.93	4.05

Suppose the emission regulation level is set to that of 1990. The Republic of Korea will have the burden of reducing 44.9 per cent, 57.5 per cent of total CO_2 emission on the year 2000, 2010 respectively. The following identity relationship holds among per cent changes in GNP (Y), Energy (E) and CO_2 Emissions (C),

$$\%Y = \frac{\%Y}{\%E} \ \frac{\%E}{\%C} \ \%C = E_{YE} \ E_{EC} \ \%C,$$

where E_{YE} = energy elasticity of GNP and
E_{EC} = CO_2 emission elasticity of energy consumption.

If energy price(P_E) is equal to marginal product of energy(MP_E), then we have the following equations

$$S_E = \frac{P_E \ E}{Y} = \frac{MP_E \ E}{Y} = \frac{dY}{dE} \ \frac{E}{Y} = E_{YE},$$

where S_E = energy ratio in GNP
P_E = price of energy
MP_E = marginal product of energy,

Now energy ratio in GNP(S_E) can be measured by the ratio of expenditure for energy of GNP, which is 0.0875 in 1990. If energy price(P_E) is fixed, then the energy ratio in GNP changes proportionally with energy intensity(E/Y). Since energy intensity in 2000, 0.72, is projected to be almost the same as the energy intensity in 1990, 0.71, the energy ratio in GNP in 2000 can be expected to be close to the level of energy ratio in GNP in 1990. Also, since the energy intensity in 2010 is projected to be 0.60, reduced in a rate of 17 per cent compared to the year 2000, the energy ratio in GNP in 2000 can be expected to be about 0.072.

The CO_2 elasticity of energy consumption moves in the same direction as the technology improvement of energy use. In industrialized countries, the elasticity is expected to be around 0.5. Since the Republic of Korea relies more on fossil fuels and has less developed technology than industrialized countries, the elasticity must be higher than 0.5. In this paper, the CO_2 elasticity of energy consumption is assumed to be 1 and 0.5. Based on; the BAU scenario, then, CO_2 emissions are required to be reduced at a rate of 44.9 per cent in the year 2000 and at a rate of 56.3 per cent in the year 2010. As a result, the Republic of Korea will have the burden of reducing GNP growth rates by 3.93 per cent point and 4.05 per cent point in the year 2000 and 2010, respectively.

4. Response Strategies of the Republic of Korea

The present energy policy of the Republic of Korea has the stated objective of seeking to secure balance and harmony among the three targets; economic development, environmental protection, and energy supply security. Since the economy is still in process of development, however, efforts to achieve the goal have been hindered by a range of obstacles including: the higher priority of economic development over environmental protection; the scarcity of financial resources necessary for establishing environmentally-sound energy systems; and lack of sufficient technologies.

Further economic development of the country will necessitate larger energy inputs in the future. For it to make the necessary transition to a less carbon-intensive future, there is a need for broad participation from the government, private sector, industry, and other economic sectors. The following are the initial areas for practical action programmes: improved energy conservation and efficiency; greater use of cleaner energy sources. First, in light of the heavy reliance of the Republic of Korea on energy imports and its energy intensive industrial structure, energy conservation and energy efficiency improvement should be the foremost priority. Since the country imports 73 per cent of its primary energy requirement (excluding nuclear power), the Government has placed a great emphasis on energy conservation and efficiency improvement. For example, in 1979 the Government established the Korea Energy Management Corporation (KEMCO) to carry out detailed policies for more efficient energy conservation, including energy audits, technical assistance, training and education and information services and publications. However, current energy conservation policies and programmes in the country have largely concentrated on well-known information and education activities, technology standards and improvements, and general rules and regulations. Advanced concepts, such as integrated resource planning, and more aggressive policies, such as market-based energy pricing, are not widely known or have not been fully adopted. In order to pursue energy conservation policies in these directions, the country has a strong need for systematic planning, measurement, and evaluation tools and procedures to clearly assess the merits, cost-benefits, and impacts of alternative programmes.

Second, the energy structure should be redesigned to an environmentally-sound energy system. For example, policies aimed at replacing more carbon-intensive fuels for cleaner sources should focus on broadening the use of natural gas and nuclear power. With the rise of environmental concerns, natural gas will come to play a significant role in the international energy market in the future. The supply of natural gas, distributed over many nations across the world, appears quite secure. In the Republic of Korea, a number of opportunities exist for substituting natural gas for oil and coal in space heating, power generation and industrial use.

Although it started to use natural gas in the form of LNG in 1987, the nation still faces a number of constraints in expanding its natural gas use due to the insufficient construction of the supply infrastructure. According to the government plans, the construction of a nationwide pipeline network will be completed in 1996. The completion of the necessary distribution network should facilitate the further incorporation of natural gas into the Korean fuel mix.

A heavier reliance on nuclear energy can also contribute to the reduction of greenhouse gas emissions, particularly in power generation. There are already 12 nuclear power plants in the Republic of Korea. However, recently the expansion of nuclear power generation has faced serious problems, such as plant siting, public reluctance towards nuclear sources and, most important, waste disposal problems. These issues stem both from safety concerns and constraints on available land. Unless these problems are overcome, the use of nuclear in the Republic of Korea will be constrained in coming years.

5.   Conclusion

According to the analysis in this paper, if no drastic measures are taken i.e. business as usual scenario – Korean energy consumption is projected to grow at an annual rate of 6.7 per cent in the period of 1991-2000 and 3.6 per cent in the period of 2000-2010. CO_2 emissions are projected to grow at an annual rate of 6.1 per cent in the period of 1991-2000 and 2.6 per cent in the period of 2000-2010. If the emission regulation is set to that of 1990, required GNP growth rate reduction is estimated to be 3.9 per cent in 2000 and 4.1 per cent in 2010.

The current Climate Treaty stipulates minimum requirements for the preparation of response strategies to combat climate change and the Republic of Korea, currently regarded as a developing country, may not be greatly influenced by the Convention in the short term. However, the country needs to prepare for the Convention itself and the subsequent development. In this respect, the basic position of the energy and environmental policy should be 'No Regret Strategy'. Some initial areas for practical action programmes for No Regret Strategy are improved energy conservation and efficiency and greater use of cleaner energy sources. In order to move in these directions, there should be broad participation from the government, private sector, industry, and other economic sectors.

Table 3.5 Summary of Major Energy Conservation Programmes in the Republic of Korea

Programmes	Purposes	Activities and Budget Support (Won)

Financial support for R & D on energy efficient technologies (ESTs)	* Expansion of support size and strength support system
	  – Support for R&D on ESTs
	  – Enhancement of research efficiency in energy conservation-related research institutes
	* Sustained support for commercialization of ESTs
	  – 100 core ESTs in industrial sector
		* Support for industry in ESTs
		  – Subsidies, 5.7 billion; loans, 0.1 billion
		  – Government finance, 1.2 billion
		  – Fund for oil industry, 4.26 billion
		  – KEPCO fund, 3.3 billion
		* Establishment of "Supporting Center for Energy-Resource Technology Development"
		* Completion of survey on market demand and selection of 100 ESTs

Establishment of 5-year conservation plan in energy-intensive industries	* Special management for 194 factories that cover 60 per cent of total energy demand in industry and each use 20,000 TOE per year
	  – 5-year plan for conservation (92-96)
	  – Efficiency targets for major items (92-96)
		* Five-year plan includes:
		  – Investment, 138.5 billion
		  – Saved energy, 2.9 million TOE/yr
		  – Saved money, 331.6 billion/yr
		  – Announcement of annual efficiency improvement targets for 161 items that consume over 1,000 TOE/yr
		* Free diagnostic analysis for 500 companies
		  – Special management for 4 types of equipment (60)
		  – Technical guides (60)
		  – Field training for energy managers (374)

Implementation of a special plan to improve the efficiency of boilers and electric motors	* Support for R&D on:
	  – Higher efficiency electric motors (10 per cent savings)
	  – Efficient boilers (30 per cent savings)
	* Establishing minimum and target efficiency standards
		* Selection of 10 target techniques that have high conservation potential
		* Conducting a survey to determine the current levels of efficiency and production of electric motors

Expansion of diagnosis and training in conservation techniques and supply of technical informaiton	* Providing incentives for investment in diagnosis industry (energy saving companies) by offering: – Free diagnosis for 250 small and medium-sized companies	* Intensive management by appointed energy specialists for 250 small and medium-sized companies
Use of pricing mechanisms for demand-side management	* Considering a flexible pricing system for oil to maintain an appropriate price level to reduce oil consumption * Rationalization of price system between high-sulfur oil and low-sulfur oil	* Maintenance of a low price for residential B-C and LPG * Study on the implementation of a flexible oil pricing system

Reduction of growth rate in electricity peak demand by 10 per cent	* Improvements in tariff structure
	* Obligation to install ice storage systems and gas heating and cooling systems
		* Interruptible rate
		  – Discount in demand charge
		  – Discount = (peak of the month – contracted peak) x discount unit x days
		* Adjusted rate
		  – Discount in demand charge
		* Increase prices
		  – For 1991, by 4.9 per cent
		  – For 1992, by 6.5 per cent
		* Gas heating and cooling system
		  – Buildings, 1,265
		  – Capacity, 375,000 RT
		  – Electricity substitution, 168 MW
		* Ice storage system
		  – Mandatory installation in new or reconstructed buildings starting 12/1/92
		* Financial support for manufacturers that produce ice storage systems
		* Tax reductions for building owners
		* Enhancement of maximum level of subsidy by KEPCO (9/92) – 55 million Won to 100 MW

Table 3.5 Summary of Major Energy Conservation Programmes in the Republic of Korea *(Continued)*

Programmes	Purposes	Activities and Budget Support (Won)

Expansion of distribution of high-efficiency appliances	* Settlement of energy labeling scheme in early stage	* Implementation of energy labeling scheme; recommended items include:
		  – Refrigerators, vehicles (9/1/92)
		  – Lighting (10/1/92)
		  – Air conditioners (1/1/93)

Improvement of building energy efficiency	* Financial assistance for building renovations to incorporate energy efficient facilities * Establishing energy efficiency standards by sector for the design of shower rooms, swimming pools, building envelopes, and mechanical and electrical equipment	* Conducting energy audits for 118 high-energy consumption buildings (7-8/92) * Supporting research by the Korea Institute of Energy Research (KIER)

Implementation of Energy Impact Assessment	* Drive for conservation through prior assessments for large public works projects
	* Guidance for energy efficient buildings in the public sector
		* Institutionalization of EIA:
		  – Law (9/92)
		  – Assessment institutions (9/12/92)
		  – Assessment committee (12/92)
		* Prior assessments for 43 governmental buildings

Expansion of use of energy-efficient equipment in the public sector	* Expansion of distribution of fluorescent lamps	
Mandatory labeling programme for fuel efficiency	* Apply mandatory labeling to all passenger cars	* Started labelling for domestic and imported passenger cars
Expansion of use of smaller cars	* Differentiate car ownership taxes and operating costs according to engine size or vehicle weight	* Levied taxes on purchase and annual road taxes based on engine size

Promotion of energy conservation during vehicle operation	* Encourage voluntary fuel-saving driving and regular maintenance
	* Encourage car pooling and curbing of cars one day in 10
	* Incentives and advertising programmes to promote car pooling
	* Intensification of fuel economy education by distributing "Fuel Economy Guide"
	* Energy conservation through the efficient freight system
	* Mitigation of traffic congestion in existing oil transport channel by constructing the metropolitan oil pipeline system
	* Accelerate the construction of the south-to-north oil pipeline system
	  – Construct 900 km oil pipe route
	  – Construct 2 oil storage terminals
		* Mandatory curbs on car use one day in 10 for all public service personnel (2/1/92)
		* Voluntary curbs on car use for the private sector (after 3/92)
		* Distribution of 50,000 booklets, "Guide to Fuel Economy and Fuel-Saving Driving"
		* Complete construction of the metropolitan oil pipeline system (12/92)
		* Procure equipment and materials for 256 km oil pipe route
		* Purchase site of oil storage terminal

Public relations programme	* Strengthen PR programme by emphasizing appropriate issues for each season
	* Encourage use of mass transit
	* Designation of November as "Conservation Month"
	* Development of texts for education
		* Establishment of PR programmes for summer and winter:
		  – Street campaign
		  – Guide and information to keep heating and cooling systems set at the levels the government recommends for buildings
		  – Education in efficient use of heating and cooling equipment
		  – Organization of seminars and conferences on conservation
		* Mass media campaigns
		  – Television, 1,189 messages
		  – Raido, 506 messages
		  – Newspapers and magazines, 1,124 messages

	Table 3.5 Summary of Major Energy Conservation Programmes in the Republic of Korea *(Continued)*	
Programmes	*Purposes*	*Activities and Budget Support (Won)*

		* Exhibitions and conferences
		    – Public rally (4,000 attended)
		    – Governmental award for contributors to conservation
		    – Exhibition programme for energy-efficient equipment
		    – Workshops and conferences
		* Development of text for primary schools and high schools

Source: KEEI		

56

CHAPTER VI

ENERGY DEVELOPMENT AND MANAGEMENT IN SRI LANKA*

1.0 Introduction

Sri Lanka is an lsland of approximately 65,000 square km and a population of seventeen million. The Country hence is characterized by a relatively high population density. The per capita GNP in 1992 was 494 US$ [1]. Sri Lanka is not endowed with any commercially exploitable fossil fuel resources, but due to its central hills and favorable weather conditions, has a good potential for hydro power generation. The country being in the tropical belt has abundant solar insolation and due to its geographic location is exposed to both South West and North East monsoonal winds. There is no significant shortage of Biomass resources and other agricultural residues in the country at present making it the main source of energy.

Due to these features, the Sri Lankan economy is somewhat different to the rest of the developing countries. The issues related to population, energy needs, human resources and environment resource constraints however are same as for other small developing countries in the region.

The per capita energy consumption inclusive of all forms of energy is of the order of 0.36 TOE per annum (TOE = 41.84 GJ). This is below the world standards but is comparable with the other developing countries in the region. The growth oriented development path, dictates that the increasing population and improved living standards needs increasing inputs of goods and services, and increasing needs of goods and services necessitates a growth in the economy and the growth in the economy ultimately leads to the growth in all perceivable inputs to the economy such as natural resources, human resources, indigenous and imported energy. This can no longer be rationally supported as in many countries in the region. As such the challenge today in all developing countries and Sri Lanka in particular is in sustained economic growth, managing within limits the exploitation of natural, and human resources, hopefully resulting in the preservation of the natural environment which sustains the life on earth.

Sri Lanka though an Island, cannot be in isolation with regard to the conservation of the environment even within the now fashionable parameters of sustainable growth. We have been and will be dependent, very significantly, on imported fossil fuels in whatever form, for the sustenance of the economy. This brings Sri Lanka into close contact with constraints in the exploration and utilisation of fossil fuels. It is also true, even if our contribution as a small country in the region, of the effluents emitted by burning fossil fuels is insignificant, we have to be mindful of immediate local impacts of fossil fuel usage, besides conservation of the regional and global environment by curtailing effluent discharges in the process of burning fossil fuels.

The environmental consequences of large dams and reservoirs constructed for our hydro power generation cannot also be disregarded. Further more these large reservoirs are also concentrated in the central hilly region and the effects on the surrounding areas is also of great concern. The inundation of large areas of productive and most fertile land for hydro power generation alone can no longer be pursued.

In the above context Sri Lanka is concerned about the preservation of the environment for the future generation, while the energy needs of the present population for its very sustenance cannot be denied. With this in view the energy planner in Sri Lanka has a multiple role to meet the energy demand, at least cost, to the economy and environment.

2.0 The Economy

The economy has had a moderate growth in the past five years and is expected to grow on the average of about 6 per cent in the next five years. The rapid industrialisation envisaged is to be the leading factor in the economy. This is quite evident from Table 2.1.

This industry-led economy is expected to improve the economy further. The Government has forecast that at least in the medium term (next five years), the current account deficit should reduce to a manageable level.

The main thrust of the economic reforms undertaken by the Government in the immediate past with the prime objective of a sustained growth is as follows [2].

* Presented by W.J.L.S.Fernando, Deputy General Manager (Generation Planning), Ceylon Electricity Board at the Expert Group Meeting Preparatory to the First Session of the Committee an Environment and Sustainable Development, Bangkok, 30 September - 2 October 1993.

Table 2.1. Growth Rates of GDP by Major Sectors at Constant prices

			Percentages				Average
	1991	1992	1993	1994	1995	1996	92/96
1. Plantation Agriculture	-5.7	-5.6	5.3	3.4	1.0	2.2	1.2
2. Other Agriculture	5.5	1.6	3.1	4.7	3.2	3.3	3.2
3. Mining, Manuf. & Constr.	5.0	7.1	6.8	8.8	8.8	8.9	8.3
4. Services	6.1	5.9	5.5	6.5	7.2	7.7	6.6
5. GDP	4.8	4.5	5.3	6.4	6.7	6.9	6.0
GDP at (1991) Constant Market prices: RS.Billion	375.3	392.2	413.1	439.8	469.0	500.9	–
GDP deflator (%)	11.3	10.9	8.1	6.0	5.0	5.0	7.0

Source: Public Investment 1992 - 1996, Dept. of National Planning, Sri Lanka [2]

1) Achieving macro economic stability.

2) Improving the effectiveness of welfare programmes.

3) Creating an environment suitable for private sector and export led development.

The economic policy as envisaged by the Government in the medium term will continue to focus on structural adjustment and stabilisation. It is also forecast that the recovery of the economies of the developed world, there will be a positive influence on Sri Lanka's export earnings, tourism and migrant worker remittances among others.

It is clearly evident that with rapid industrialisation in the medium term, the demand for energy will undoubtedly increase.

3.0 Energy situation in the Country

The main source of energy in Sri Lanka is biomass. The other domestic resource is hydro. The domestic indigenous resources constitute more than 80 per cent of the energy supply in the Country. This however does not mean the contribution in the form of useful energy it has the same importance as the shares indicate. A significant component of the biomass consumption is the domestic sector with very low efficiency of conversion, mainly for cooking. As such, the contribution of biomass as an energy source in the national economic activity is marginal. This does not however mean that the importance of biomass as the main source of the energy in the Country should be disregarded.

Sri Lanka has no proven reserves of fossil fuels. A small quantity of peat has been located in the vast extent of marshy lands to the North of Colombo. A recent feasibility Study [3] has indicated that the quality and extent of the reserve could not prove to be commercially viable for extraction and use in power generation.

As stated earlier, Fuelwood and other biomass including agro residues and hydro electricity are the main indigenous primary sources of energy supply in the country. Crude oil is the main primary energy import. Coal and refined petroleum products such as LPG, Diesel and Kerosene are also imported in small quantities.

The gross and the useful energy supply is shown in fig. 3.1 Hydro Electricity supply has been adjusted to reflect the energy input to a thermal plant to produce the equivalent amount of electricity (oil replacement Value: $1 kWh = 0.24 \times 10^{-3}$ TOE).

The development of hydro resources and the import of other forms of commercial energy namely oil causes a severe burden on the economy in terms of utilisation of foreign exchange. Even if the development of hydro resources for electricity generation is on the decline, in the medium to long term, fossil fuel requirements to meet the energy demand will significantly increase. Therefore, there will be an increased dependence on imported energy supplies in the future.

As stated earlier, the per capita energy consumption including traditional non commercial energy sources in the country is 0.36 TOE. A slight decline of the energy consumption is seen over the past few years. The change is due to a reduced estimate of fuel wood consumption. The consumption of energy by sectors of all forms of enegy and commercial energy is shown in Fig. 3.2.

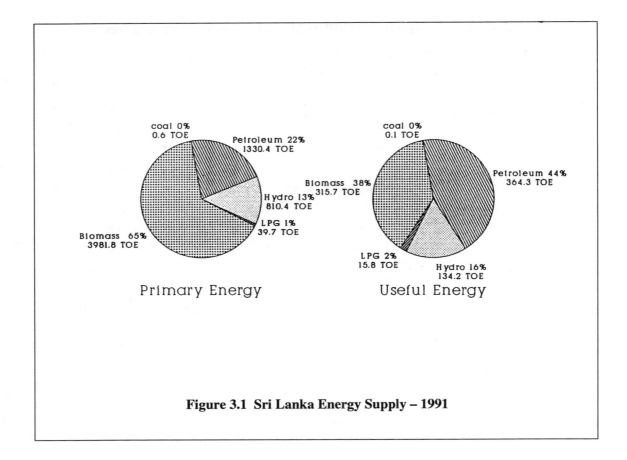

Figure 3.1 Sri Lanka Energy Supply – 1991

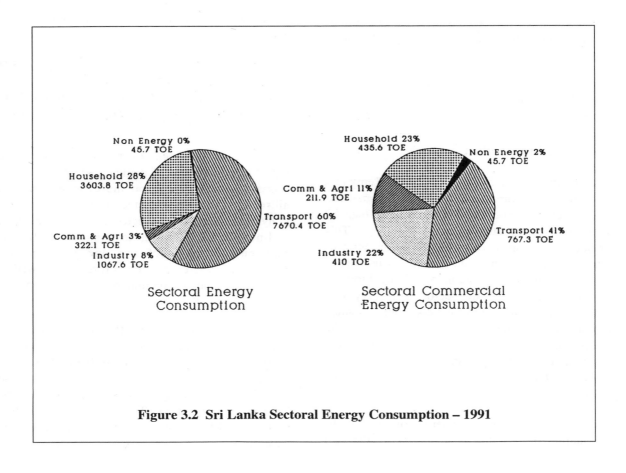

Figure 3.2 Sri Lanka Sectoral Energy Consumption – 1991

4.0 National Energy Policy

The development of the economy of Sri Lanka and improvement of the living conditions of its people will require the use of more energy in the coming years. The expected rate of increase in electricity and petroleum products will be more than that for biomass, as new industry develop and more households are electrified. Hence, a more rational approach for development of the energy sector is needed. With this in view, the following National Energy Policy was adopted by the government.

1) Providing the basic human energy needs.

2) Choosing the optimum mix of energy resources to meet the requirements at the minimum cost to the national economy.

3) Optimisation of available energy resources (hydro electric, biomass, solar, wind and petroleum) to promote socio-economic development.

4) Conserving energy resources and eliminating wasteful consumption in the production of energy and the use of energy.

5) Developing and managing of forest and non-forest wood fuel resources.

6) Reducing dependence on foreign energy resources and diversifying the sources of energy imports.

7) Adopting a pricing policy which enables the financing of energy sector development.

8) Ensuring continuity of energy supply and price stability.

9) Establishing the capability to develop and manage the energy sector.

5.0 Overall Energy Supply/Demand and Conservation Measures

With the moderate economic growth in the past few years the dependence of GDP growth on the growth of energy has not increased in the recent past. This phenomenon is favorable to a country such as Sri Lanka and may have been due to conservation measures. However with rapid industrialisation forecast in the near future one could no longer expect this trend to be sustained.

The expected demand of petroleum products in the current decade is to increase at an average rate 2 per cent per year while the electricity demand is expected to grow at the rate of 7 per cent on the average per year. The biomass consumption is expected to rise at about 1.5 per cent per annum on the average. The aggregate annual average rate of growth of energy is projected to be 2.5 per cent in the first half of current decade and is expected to rise to about 3.2 per cent in the latter half [4]. The relative composition of demand growth of petroleum biomass and electricity poses widely differing issues related to energy policy, energy pricing and environment.

This projected energy demand growth in the next ten years suggests the following policy options about supply and demand [5].

Energy Source	Supply Options	Demand Management Options
Fuelwood & Biomass	Improve Supply to Urban & Estate areas	Efficient use for domestic cooking
	Wood Lots to be raised by fuel-wood using industries	Efficient use in industry
	Raise fuel wood plantations	
	Develop Use of Agro-wastes	
Petroleum	Improve and optimise refinery yield to meet the demand	Efficient use in industry
	Develop Storage facilities	Optimise use in transport
Electricity	Development of thermal and hydro plants	Conservation in all sectors
	Develop distribution network	Demand Side Management
	Reduce Transmission and Distribution losses	Compact Fluorescent Lamps.
	Develop alternative Eg: Wind	T and D loss reduction
		Improvement of Load factor

5.1 Fuelwood and Biomass

The data on demand for fuelwood and agro-waste, which comprise the major component of biomass in the country, is not so accurate as the other sources. There have been ad-hoc surveys carried out since late seventies to estimate the biomass consumption in the country. The actual and projected demand of biomass as estimated by Perera [4] using available survey data are given in table 5.1.

Table 5.1 Demand on Biomass

Year	Demand *Primary Energy 000 TOE*
1989	4642
1995	5154
2000	5610

Source: Energy Status of Sri Lanka, [4]

Out of this total demand, 83 per cent is used for household cooking purposes by approximately 93 per cent of the population. The balance 17 per cent of the total is used in the industry. Almost 33 per cent of the industrial consumption of firewood is in the tea industry while the rest is consumed in Brick and Tile industry (13 per cent), Coconut industry (11 per cent) Tobacco industry (10 per cent), Rubber industry (6 per cent) and the rest in other sectors including the commercial sector.

There is no apparent shortage of biomass in the country as a whole. However, there are regional deficits which makes firewood an expensive fuel in the urban areas. Table 5.2 reproduced from Perera [4] indicates that there is bound to be fuelwood shortage from the year 2000.

With the impending fuelwood shortages at the turn of the century and with hardly any shift in demand for any other fuel (even in the year 2000 – 92 per cent of the households is estimated to be cooking using fuelwood) considerable effort has to be made in meeting the fuelwood demand from the year 2000. The Forestry Master Plan has proposed an extensive forest and fuelwood management programme which will enable a supply of 10 per cent higher than what is estimated. Hence the availability of fuelwood to meet the demand is expected to be extended to the year 2020.

The National Fuelwood Conservation Programme [6] handled by the Ministry of Power and Energy and the Ceylon Electricity Board is of importance in reducing the demand for fuelwood especially in the domestic sector. The urban and rural programmes are distinctly different in approach.

Table 5.2. Availability of and Demand for Fuelwood (1984-2020)

Source	1984-1985[1]	1986-1990	1000	1991-1995	tons/a[2]	1996-2000	2001-2010	2011-2020
Available								
Non-Forest Wood[3] resources	7 100	9 363		9 335		9 068	9 062	10 091
Natural forests[4]	1 770	1 273		1 205		1 137	1 035	760
Forest industry residues	250	271		319		361	435	533
Existing forest Plantations	50	468		770		393	335	273
Total Availability	9 170	11 375		11 629		10 959	10 867	11 657
Demand								
Households	7 900	8 360		8 957		9 472	9 906	10 139
Industry	1 130	1 160		1 220		1 270	1 280	1 260
Poles, posts	140	155		182		208	249	315
Total demand	9 170	9 675		10 359		10 950	11 435	11 714
BALANCE		1 700		1 270		9	−568	−57

Source: Forestry Master Plan for Sri Lanka, 1986

[1]   Estimated supply/use

[2]   Conversion factor: 1 m³ fuelwood/fuelwood equivalent = 0.7 ton fuelwood/biomass 918 per cent moist.)

[3]   Includes rubber and coconut log volumes which are not used in mechanical wood industries as well as the log volumes from homestead gardens, which the owners are using as fuel wood.

[4]   Assumes 10000 ha/a of Dry Zone natural forest clearings for agricultural development

It is envisaged that by the end of 1993, under the rural programme (which commenced in 1984) 500,000 stoves will be installed. The rural stove which is installed insitu is part subsidised. This target will be easily met by end of this year.

In the urban programme which commenced in 1987, the stove is commercially produced and sold in the market. By the year 1990, the production capability of the urban stoves has reached a target of 100,000 stoves per year.

The most recent estimates indicate that approximately 700,000 improved efficient stoves are being used both in the urban and rural sectors. The recent surveys carried out indicate that in more than 85 per cent of households in the project, 25-30 per cent of fuelwood is saved. This is a very encouraging indicator and the Government through the Ministry of Energy Conservation is actively supporting this programme.

5.2 Petroleum

Most of the domestic and foreign bunker needs of petroleum products are met by the oil refinery owned by the Ceylon Petroleum Corporation – a statutory body. There is however a shortfall of domestic production in petrol, diesel, fuel oil, aviation turbine fuel (Avtur) and LPG. The total imports/exports and foreign bunker sale of petroleum products and crude is given in Table 5.3.

Table 5.3: Petroleum imports/exports, bunker and domestic consumption – 1992

Product	Imports ’000MT	Exports ’000MT	Bunkers ’000MT	Domestic Consumption ’000MT
Crude	1 303.553			
Petrol	47.781			164.519
Diesel	420.129		46.927	641.38
Fuel oil	70.685	5.706	237.096	152.189
Kerosene				189.373
Avtur	123.437		49.784	81.398
Residual				39.19
LPG	31.382			44.93

Source: Ceylon Electricity Board [7]

Energy conservation programmes in the industrial and commercial sectors were initiated in the year 1983. Since then, over 150 engineers and technicians have been trained in energy conservation techniques. Most of the energy conservation efforts have been successful in the industrial and commercial sectors and there has been a significant impact in savings of petroleum products used for process heat generation. This conservation effort perhaps is partially responsible for the decline in the energy dependence of the GDP Growth. Perera [4] has made an attempt to estimate the possible savings due to energy conservation measures in the industrial and transport sector. Table 5.4 gives the theoretical and practical energy savings potential in major industries in Sri Lanka.

Table 5.4. Theoretical and Practical Energy Saving Potential in Major Industries of Sri Lanka

Industry	Energy Consumption	Theoretical Maximum Saving		Practically Possible Saving	
	TOE	%	TOE	%	TOE
Cement	62 400	38	23 914	20 to 25	12 480 to 15 600
Ceramics	22 393	68	15 135	30 to 40	6 718 to 8 957
Paper	20 572	100	20 572	20 to 25	4 114 to 5 143
Tyre	6 008	46	2 756	30 to 40	1 802 to 2 403
Steel	7 151	29	2 076	25 to 30	1 788 to 2 145
Glass	1 706	–	–	–	–
Total	120 229		64 453		26 902 to 34 248

Perera [4] has also made an estimate of possible savings of petroleum products in the transport sector and estimates that approximately 10 per cent could be saved.

It shall specifically be mentioned that energy conservation measures in the transport sector, has many other influencing factors, such as road condition, passenger preferences and convenience, cost of vehicles and maintenance. However more emphasis is necessary in the demand management and conservation in transport sector as this alone constitute for about 55 per cent of oil imports in the country.

Some of the conservation measures in the transport sector include regular vehicle maintenance, design and maintenance of energy efficient road network, railway electrification, use of larger busses in selected routes and an overall national transport policy directed towards energy efficiency among other considerations.

5.3 Electricity

Electricity sector planning and supply options are dealt with separately and more extensively. It would be incomplete if no mention is made of the conservation potential of electrical energy. The electrical system losses including transmission and distribution losses were in the region of 20-23 per cent in the early eighties.

A World Bank assisted loss reduction programme known as DERP (Distribution Extension and Rehabilitation Project) was launched in the mid eighties to bring down the T and D losses to a manageable 12 per cent.

The system load factor is around 55 per cent. This is to be progressively improved to 58 per cent by the year 2007. Due to this too losses are expected to be lower in the long run. The potential saving shown in Table 5.4 above includes electricity savings due to improved efficiency of usage in the industry and commercial sector as well.

Demand Side Management (DSM) has a major role to play in the conservation of electrical energy. The demand for electricity could be reduced by improving the efficiency of the end use devices or equipment. The lighting, a major end use, consuming about 20 per cent of the total demand has a high potential for conservation. The incandescent lamps (IL) are widely used for lighting. Compact Fluorescent Lamp (CFL) which is proven to be energy efficient and having a long life is a good substitute for IL.

The impact of the replacement of 60W incandescent lamps with 11W CFLs is given below.

Data used for the calculation

		IL	CFL
Wattage (W)	=	60	11
Life time (hrs)	=	1000	8000
Cost (RS.)	=	25	750

No of lamps to be replaced = 1,000,000

Assumptions

1. Lamps are operated 4 hrs a day during night peak.

2. All the lamps are switched on at the sametime.

It could be seen that with the implementation of a DSM programme, the demand for power during peak hours could be reduced by 49 MW and energy by 72 GWh. That is about 6 per cent reduction of the peak demand of 1994. It shall be cautioned that since no diversity is allowed in this calculation, the estimated capacity saving during peak time can be different.

The cost saving to CEB is calculated with the following assumptions:

The detailed calculations are given in Annex A1.

1. The construction and operation of new GTs are avoided due to the above replacement.

2. The costs of CFLs are to be borne by the CEB.

The cost comparison too is given in Annex A1.

Over the life time of CFL, saving on construction and operation of GTs are 2408.56 M.Rs. and revenue loss to CEB is 1306.80 M.Rs. Since the costs of CFLs are to be borne by the CEB (ie 750 M.Rs.), net saving to CEB is 351.76 M.Rs.

A pilot project is presently being carried out by CEB with the assistance of the Energy Conservation Fund and University of Moratuwa to investigate the feasibility and the economics of the use of energy efficient lighting in households. This study is a prelude to a large project funded by the World Bank.

5.4 Overall Energy Conservation Targets

The overall achievable energy conservation targets in biomass, oil and electricity is given in Table 5.5. These does not include the DSM measures above using compact fluorescent lamps in the Electricity Sector. These targets, if achieved, can be very favorable to a country like Sri Lanka.

Table 5.5. Effect of Conservation Measures on Projected Overall Primary Energy Demand

	Million Tonnes of Oil Equivalent				
	1989	1995		2000	
		Without conservation	With conservation	Without conservation	With conservation
Biomass	4.35	4.66	3.79	4.93	4.06
Petroleum oil	1.23	1.50	1.44	2.13	2.02
Hydro-elect.	0.67	0.93	0.90	1.05	1.05
Total	6.25	7.09	6.13	8.11	7.13

Source: Energy Status of Sri Lanka [4]

6.0 Electricity Sector: Planning, Environmental Impacts and Demand Side Management

Ceylon Electricity Board (CEB) is the statutory body with the responsibility for the Generation, transmission and most of the Distribution of electrical power in Sri Lanka. Presently the CEB operates 12 major hydro power stations and 3 thermal power stations feeding the national grid. The total installed capacity of all the power stations owned and operated by the CEB is 1330 MW. In the year 1992, the peak demand recorded was 742 MW and the electrical energy generated was 3540 GWh. Out of this, 640 GWh of electrical energy was supplied by the thermal plants.

At present, the electric power generating system of Sri Lanka is predominantly based on hydro energy. Eighty per cent of the total installed system capacity is from the 12 major hydro power stations. In the year 1990, 99.8 per cent of the energy demand of the system was met by the hydro plants. However in the years 1991 and 1992, only 92.3 per cent and 82 per cent of the system energy demand was met by hydro energy respectively. Table 6.1 shows the hydro and thermal share for the past few years.

Table 6.1. Electricity Generation 1981 - 1992

Year	Hydro Generation		Thermal Generation		Total
	GWh	%	GWh	%	GWh
1981	1 572	84.0	300	16.0	1 872
1982	1 608	77.8	458	22.2	2 066
1983	1 217	57.6	897	42.4	2 114
1984	2 091	92.5	170	7.5	2 261
1985	2 395	97.2	69	2.8	2 464
1986	2 645	99.7	7	0.3	2 652
1987	2 177	80.4	530	19.6	2 707
1988	2 597	92.8	202	7.2	2 799
1989	2 801	98.0	57	2.0	2 858
1990	3 144	99.8	5	0.2	3 149
1991	3 116	92.3	260	7.7	3 376
1992	2 900	81.9	640	18.1	3 540

Source: Ceylon Electricity Board [8]

The total installed capacity of hydro plants, excluding small hydro plants, is 1115 MW. These plants are located in two major river basins namely Kelani and Mahaweli. The recently commissioned Samanalawewa hydro electric power plant is located in the Walawa ganga river basin. Details of plant capacity and the long term average expected generation are given in Table 6.2.

The total installed capacity of the thermal plants is 216 MW. Details are given in Table 6.3.

6.1 Power System Planning

The need for the long term plan for a power system is enumerated below; [9]

1. the long lead time and long period of utilization of electric power facility,

2. the electric power sector is highly capital intensive and often constitute the largest single investor in a country. The expenditure in the power sector and the tariff structure from such investments affect the national economy,

3. the environmental impacts of power sector development and the necessary policy to minimise the environmental impacts at an affordable cost over time,

4. the long term requirement of human resource development and the need for technical education and training.

6.1.1 Generation Expansion Planning Tools

The main planning tool in the analysis of the generation expansion is the "Wien Automatic System Planning" Package (WASP III). The original version of WASP-III was first developed by the Tennessee Valley Authority (TVA) and Oak Ridge National Laboratory in the USA. International Atomic Energy Agency (IAEA) has maintained a continuous programme of work on WASP-III planning package and trained personnel in the use of this tool. Sri Lanka too has received assistance from IAEA in the form of training of personnel, software support and exchange of information through seminars and workshops.

The WASP-III computer programme is designed to find the economically optimal generation expansion policy for an electrical power generating system, with user specified constraints. The programme consists of seven modules. [9]

During the period 1986-1989, with the assistance from the Federal Republic of Germany, a project titled Masterplan for the Electricity Supply of Sri Lanka was carried out to; [9]

Table 6.2. Existing and Committed Hydro Power Plants

HYDRO PLANT	INSTALLED CAPACITY (MW) Units x capacity		AVERAGE FLOW (m^3/s)	ELECTRICITY GENERATION (Gwh/yr) Average Historic	Long Term Expected	COMMISSIONING
Laxapana (KM) Complex						
Canyon	2 x 30	60	12.7	132	162	Unit 1 Mar 1983 Unit 2 1988
Wimalasurendra	2 x 25	50	7.6	120	112	Jan 1965
Old Laxapana	3 x 8.33 2 x 12.5	50	9.1	262	286	Dec 1950 Dec 1958
New Laxapana	2 x 50	100	16.2	460	480	Unit 1 Feb 1974 Unit 2 Mar 1974
Polpitiya	2 x 37.5	75	9.1	381	417	Apr 1969
Laxapana Total Capacity		335		1 355	1 457	
Mahaweli Complex						
Victoria	3 x 70	210	57.3	588 (1985-92)	734	Unit 1 Jan 1985 Unit 2 Oct 1984
Kotmale	3 x 67	201	30.8	404	483	Unit 1 Apr 1985 Unit 2 Feb 1988 Unit 3 Feb 1988
Randenigala	2 x 61	122	75.3	260 (1987-92)	381	Jul 1986
Ukuwela	2 x 19	38	34.1	163	175	Unit 1 Jul 1976 Unit 2 Aug 1976
Bowatenna	1 x 40	40	48.2	55 (1987-91)	53	Jun 1981
Rantembe	2 x 24.5	49	94.1	131 (1990-92)	214	Jan 1990
Mahaweli Total Capacity		660			2 040	
Samanalawewa	2 x 60	120	17.3	–	288	Oct 1992
Small Hydros						
Inginiyagala	2 x 2.475 + 2 x 3.15	11	n.a.	27	n.a.	Jun 1963
Uda Walawe	3 x 2	6	n.a.	8	15	Jul 1988
Nilambe	2 x 1.6	3		11 (1990-92)		
Small Hydro Total		20				
Total excluding small hydro		1 115		3 785		

Source: Ceylon Electricity Board [8]

Note: Historic average given for 1984-92, unless stated otherwise

Table 6.3. Details of Existing Thermal Plants

Plant	Capacity Units x MW	Commissioned Unit	Commissioned Date	Remarks	Average Annual Plant factor %
Kelanitissa Gas Turbines	6 x 8 (originally 20 MW, Now derated)	1 2 3 4 5 6	Nov 80 Mar 81 Apr 81 Dec 81 Apr 82 Mar 82	To be overhauled in 2002 and 2003 3 units each year	14.0 (1983 - 1992)
Kelanitissa Steam	2 x 22 (originally 25 MW, Now derated)	1 2	Jun 62 Sep 63	Recommissioned in 1991 after rehabilitation. To be retired by end 2000	15.6 (1971 - 1992) excluding 1986 - 1989
Sapugaskanda Diesel	4 x 16 (originally 20 MW, Now derated)	1 2 3 4	May 84 May 84 Sep 84 Oct 84	To be retired by end 2003 and end 2007, 2 units each year	13.6 (1985 - 1992)

Source: Ceylon Electricity Board [8]

- inventorise the hydro-electric potential
- inventorise non-hydro options
- system operation and expansion studies

A number of programmes and tools developed during this study are presently used by CEB's planning branch staff for the development of the long-term power system of Sri Lanka. some of the useful tools developed were;

SYSIM – computerized simulation model for hydro-thermal, irrigation system of Sri Lanka to analyse the energy potential of existing and future hydro plants,

EVALS – project dimensioning and costing programme of selected candidate hydro plants,

EXTRA – computerized model for the expansion of transmission system with reliability criteria with DC load flow analysis and

SEXSI – system expansion simulation model to study the future power generation expansion sequence including uncertainty in planning.

6.1.2 Generation Expansion Planning Process

In order to develop the long term generation expansion plan, the following tasks are carried out:

- Collect data on the existing generating system
- Prepare electricity demand forecast
- Identify candidate generating projects
- Determine the optimum plant addition sequence

The long term power and energy demand forecast is updated annually based on information such as past sales growth, GDP growth rate etc. Table 6.4 shows the past electricity generation and sales figures.

The past trend analyses shows that during the last 20 years the average long term growth of electrical energy was 7.2 per cent per annum. This means, the demand for electrical energy will double every 10 years. Base case demand forecast for the period 1993 to 2012 prepared in 1992 is shown in Table 6.5

Table 6.4. Electricity generation, peak demand and their annual growth rates.

Year	Total Sales (GWh)	Per Capita sales (GWh)	Percentage of Sales by Tariff Group			
			Domestic	Industry	Commerce	Other*
1971	722	57	8.9	51.7	12.9	26.5
1972	823	65	8.7	54.4	12.0	24.9
1973	867	66	9.5	53.8	12.4	24.3
1974	907	68	9.1	54.2	13.0	23.7
1975	965	71	9.0	54.3	12.4	24.3
1976	997	73	9.5	51.8	13.5	25.2
1977	1 041	75	10.2	49.9	14.2	25.7
1978	1 166	82	10.2	50.9	13.9	25.0
1979	1 298	90	11.8	48.7	15.5	24.0
1980	1 392	94	13.7	45.0	16.0	25.3
1981	1 510	101	14.3	44.9	14.6	26.2
1982	1 694	112	15.3	43.6	15.5	25.6
1983	1 797	117	17.0	41.8	16.2	25.0
1984	1 886	121	16.8	41.3	17.0	24.9
1985	2 060	130	16.8	41.3	17.0	24.9
1986	2 232	138	16.5	41.5	17.1	24.9
1987	2 253	138	16.9	38.5	18.6	26.0
1988	2 371	143	16.5	38.1	19.5	25.9
1989	2 353	140	17.3	36.1	19.1	27.5
1990	2 608	153	19.0	34.9	16.2	29.9
1991	2 657	161	23.5	34.9	19.9	21.7
1992	2 869	165	23.1	36.7	19.6	20.6

Source: Ceylon Electricity Board [8]

* Includes bulk sales to Lanka Electricity Company and Local Authorities, some of which are gradually being taken over by CEB.

Table 6.5. Base Forecast 1993

Year	Sales (GWh)	Annual Growth rate (%)	System Losses (% of generation)	Generation (GWh)	Load Factor (%)	Peak Demand (MW)
1990	2 608	10.84	17.2	3 150	56.2	640
1991	2 742	5.14	18.8	3 377	56.3	685
1992	*2 869	5.98	19.0	*3 540	*54.5	742
1993	3 085	6.16	18.9	3 802	54.6	795
1994	3 319	7.59	18.4	4 069	55.1	843
1995	3 589	8.13	17.9	4 371	55.5	899
1996	3 886	8.28	17.4	4 706	55.9	961
1997	4 213	8.41	16.9	5 071	56.2	1 030
1998	4 567	8.40	16.5	5 469	56.5	1 105
1999	4 952	8.43	16.1	5 901	56.8	1 186
2000	5 370	8.44	15.7	6 371	57.0	1 276
2001	5 819	8.36	15.3	6 870	57.2	1 371
2002	6 308	8.40	14.9	7 412	57.4	1 474

Table 6.5. Base Forecast 1993 *(continued)*

Year	Sales (GWh)	Annual Growth rate (%)	System Losses (% of generation)	Generation (GWh)	Load Factor (%)	Peak Demand (MW)
2003	6 832	8.31	14.6	7 998	57.6	1 585
2004	7 404	8.37	14.3	8 638	57.7	1 709
2005	8 018	8.29	14.0	9 321	57.8	1 841
2006	8 688	8.36	13.7	10 068	57.9	1 985
2007	9 409	8.30	13.4	10 863	58.0	2 138
2008	10 195	8.35	13.1	11 732	58.0	2 309
2009	11 042	8.31	12.8	12 661	58.0	2 492
2010	11 966	8.37	12.6	13 693	58.0	2 695
2011	12 964	8.34	12.4	14 800	58.0	2 913
2012	14 039	8.29	12.2	15 989	58.0	3 147
2013	15 214	8.37	12.0	17 290	58.0	3 403

Source: Ceylon Electricity Board [8]

* Power cuts in April and May 1992, 1990-1992 are actuals, 1993-2013 are forecast

6.1.3 Future Generation Options

Available generation options are :

— Hydro Power

— Fossil Fuel Thermal

   — Coal

   — Medium speed Diesel

   — Low Speed Diesel

   — Gas Turbine

   — Combined Cycle

— Nuclear Thermal

— Renewables

   — Wind

   — Solar

   — Dendro etc.

In order to eliminate uneconomical options, comparison of specific generation costs, i.e. average cost of generating a kWh, over the life of the plant, was used for initial screening. These studies show that the major generating options, that may be considered for Sri Lanka to meet the future electricity demand at least cost will be mainly from a combination of hydro and conventional thermal plants.

The list of additions of new generating plants, according to the least cost plan produced during the year 1993 is given below in Table 6.6. In addition to the least cost plan, numerous sensitivity studies are carried out to check the robustness of the least cost plan by changing the basic study parameters such as load forecast, fuel price forecast, discount rate, unserved energy cost etc.

The detailed studies reveal that the share of thermal generation increases with time, which means, the burden of meeting the demand will progressively shift from hydro to thermal. From figure 6.2 it could be seen that thermal energy, which helped during the droughts in the past, will have to take over the responsibility of becoming the mainstay from the latter part of this decade.

The recommended expansion plan by the CEB based on the studies carried out by Generation Planning Branch of the CEB is then adopted by the Department of National Planning of Ministry of Policy Planning and Implementation to formulate the public investment programme. The selected electric generating facility is then considered for implementation. In Sri Lanka most of the power generating facilities are implemented through the assistance form International Lending Agencies such as World Bank, Asian Development Bank, Overseas Economic Cooperation Fund – Japan etc. Due to long time required in negotiating a loan for the implementation of these power generating facility, it is essential to negotiate the loan as soon as practicable to avoid delays in the implementation.

Table 6.6. Results of Generation Expansion Planning Studies 1993 – BASE CASE

Year	Hydro Additions	Thermal Additions	Thermal Retirements	Lolp %
1994	–	–	–	5.678
1995	–	–	–	8.182
1996	–	Gas Turbine 66 MW Diesel 40 MW (Ext.)	–	4.444
1997	–	Diesel 110 MW	–	0.726
1998	–	Diesel 40 MW	–	1.794
1999	Kukule 70 MW	–	–	3.302
2000	Upper Kotmale 123 MW	–	–	3.025
2001	–	Coal Trinco Unit 1 150 MW	KPS Oil Steam 2*22 MW	1.890
2002	–	Coal Trinco Unit 2 150 MW	Gas Turbine 3*18 MW (for refurbishment)	1.659
2003	–	Refurbished GT 3*20 MW	Gas Turbine 3*18 MW (for refurbishment)	4.698
2004	–	Refurbished GT 3*20MW Coal Trinco Unit 1 300 MW	Sapu Diesel 2*16 MW	0.418
2005	–	–	–	1.814
2006	Ging Ganga 49 MW	Gas Turbine 22 MW	–	4.045
2007	–	Coal Trinco Unit 2 300 MW	–	1.069
2008	–	Gas Turbine 66 MW	Sapu Diesel 2*16 MW	3.253
Total PV Cost upto 2013	1598.2 million US$ (76,711.9 million Rs.)			
Long Term Average) Generation Cost)	6.03 USCts/kWh (2.89 Rs./kWh)			

Source: Ceylon Electricity Board [8]

Note: Discount rate 10 per cent, Long term average generation cost calculation excludes energy contribution from existing hydro plants, plant commissioning and retirement at the beginning of the year indicated.

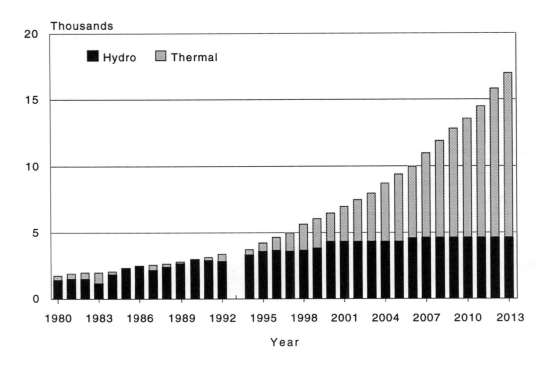

**Figure 6.2. Hydro and thermal generation
(Actual 1980-92, Expected 1994-2013)**

6.2   Environmental Impacts of Power System

The impacts of the electric power generating system on the environment and the society have been of increasing importance over the last decade. Some effects of power generating systems, present and future, that may cause impacts, both desirable and undesirable, on the environment and the society are the following:-

(a)   Gaseous emissions (CO_2, SO_x, NO_x)

(b)   Particulate emissions

(c)   Warm water discharge

(d)   Ash disposal

(e)   Land use and displacement

(e)   Employment effects

Analyses presented earlier in this report under Electricity Sector Planning considered the environmental impacts and mitigatory measures of prospective generating plants on the basis of those recommended in the relevant feasibility or pre-feasibility studies of the candidate projects.

Reference [10] reports the first attempt to investigate the environmental impacts against the costs of alternative generation plans and policies. A series of

trade-off curves were developed to identify the group of acceptable solutions (generation plans) against each of the environmental attributes. The attachment of weighting factors to the large number of attributes would be one method of developing a single representative index of the environmental impacts of power generation. Another approach that may be used is to quantify the impacts in monetary terms. A more detailed description of this study is given in Annex A2.

The impacts of the 1993 proposed (base case) generation expansion plan is investigated and reported here. Options to change the impacts by the selection of other types of plants, fuels or sites are not investigated. Such investigating measures, if considered viable or required during feasibility/pre-feasibility studies are already included in the candidate options and costs are taken in the planning studies.

6.2.1 Gaseous Effluents

Figures 6.3 (a), 6.4 (a) and 6.5 (a) show the emissions of CO_2, NO_x and SO_x if the base case plan in Table 8.1 is implemented. In the accompanying Figures 6.3 (b), 6.4 (b) and 6.5 (b), the emissions of the power sector are compared with the present and the expected emissions from the other commercial energy uses in the country.

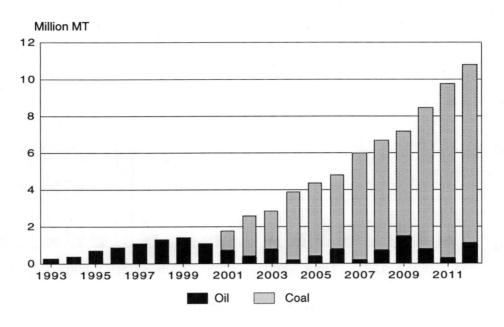

Million MT

Under average hydrological conditions

Figure 6.3(a). Expected power sector CO_2 emissions

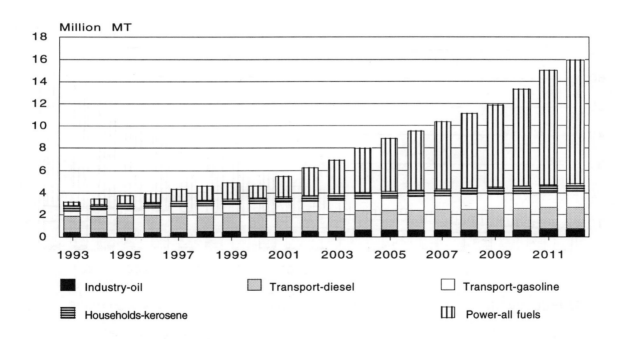

Million MT

Under average hydrological conditions
for power sector

Figure 6.3(b). CO_2 emissions by sector

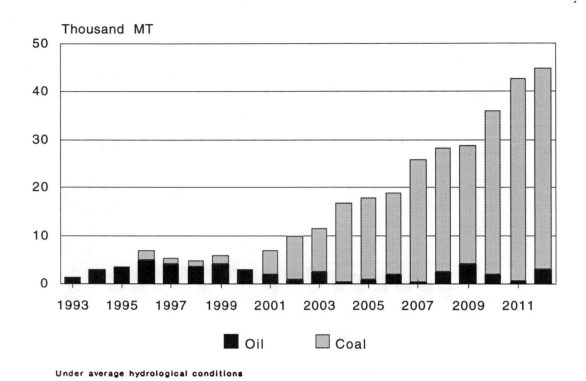

Figure 6.4(a). Power sector NO$_x$ emissions

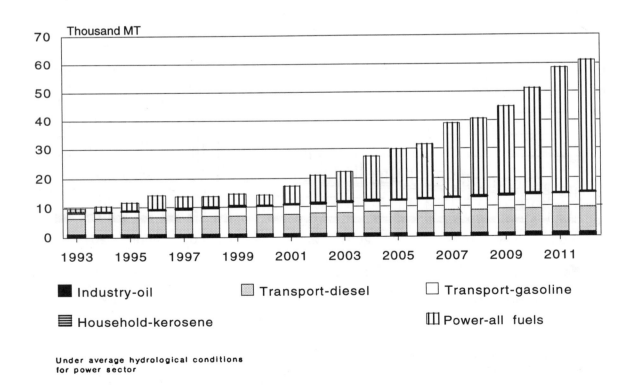

Figure 6.4(b). NO$_x$ emissions by sector

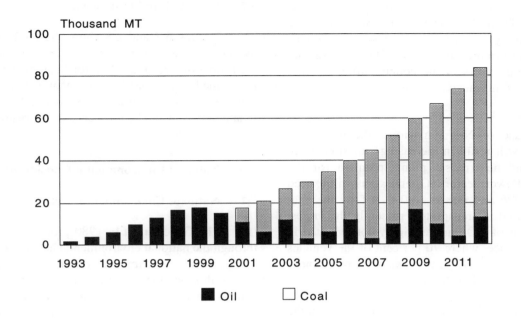

Thousand MT

■ Oil   □ Coal

Under average hydrological conditions

Figure 6.5(a). Power sector SO$_x$ emissions

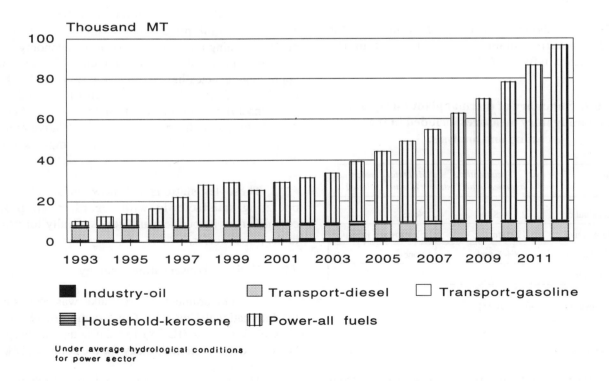

Thousand MT

■ Industry-oil   ▦ Transport-diesel   □ Transport-gasoline

▤ Household-kerosene   ⦀ Power-all fuels

Under average hydrological conditions
for power sector

Figure 6.5(b). SO$_x$ emissions by sector

72

If the Base case expansion plan is implemented, CO_2 emissions from the power sector will exceed 2 million MT/year by the year 2002. The power sector will be the dominant source of CO_2 among the commercial energy users in the country after the year 2003. Diesel used in transport is the largest producer of CO_2 at present.

In the case of NO_x emissions, the base case expansion plan will result in increasing emissions from the existing and new diesel plants planned for the medium term. Power Sector NO_x emissions will exceed 5,000 MT in 1996 and rise to 45,000 MT by the year 2012. Diesel used in transport is presently the largest single contributor of NO_x, and is highly concentrated in the Western province, where 70 per cent of the vehicles are used. Power sector NO_x emissions will exceed those of the transport sector after the year 2002.

Emissions of SO_x, most of which now occur in the transport sector, will rapidly increase with higher thermal generation. The power sector will be the largest single source of SO_x from 1996 onwards, if the Base Case expansion plan is implemented. The reason for the rapid increase of emissions in the medium term would be the increased use of diesel plants which run on high sulphur residual fuel.

Table 6.7 shows the underlying assumptions for fuel and generating plant types that result in the emissions shown in figures 6.3, 6.4 and 6.5.

Table 6.7. Summary of thermal plant fuel quality and mitigatory measures included in the costing and analysis.

Plant/Fuel	Specifications assumed in analysis	Mitigatory measures already included
Diesel/Residual oil	2.5 % S fuel	None
Coal/Bituminous coal	1% S fuel	Electrostatic Precip.
Gas turbine/diesel	0.8% S fuel	None
Combined cycle/diesel	0.8% fuel	None

Source: Ceylon Electricity Board [8]

6.2.2 Particulates

Figure 6.6 shows the particulate emissions from the power generating plants, if the base case expansion plan is implemented. After the coal plants are implemented, particulate emissions are controlled owing to limited use of plants using heavy oils.

6.2.3 Land Use and Displacement

Figures 6.7 and 6.8 show the land use for both hydroelectric and thermal plants, and the relocation of the people owing to land use for hydroelectric plants. It is assumed that the land requirement is committed at the beginning of the construction of a plant. Resettlement of inhabitants is also taken to occur at the beginning of the construction period.

7.0 National Environmental Considerations

7.1 National Environmental Act

The Government in 1980 promulgated legislation through the national environmental act No. 47 of 1980 to safeguard adverse impacts on the environment. [11] Under this act the Central Environment Authority, a central body with powers, functions and duties for the protection, management and enhancement of the environment has been established. This authority has powers inter-alia; for the regulation, maintenance and control of the quality of the environment; for prevention abatement and control of pollution. This act has further been emended in 1988 to give more power to the authority to enforce regulations as well.

Under this act regulations have been invoked in 1993 to appoint project approving agencies for all projects coming under all sectors of the economy and on various aspects of the environment. These project approving agencies have extensive powers under the act to reject or direct mitigatory measures to be undertaken for conservation of the environment. The Act also provide for public hearing and other legislative remedies in case of projects having adverse impacts on the environment.

These legislations are now enforced and are operating satisfactorily with regard to all projects coming under its purview and very specially for projects in the power sector.

7.2 National conservation strategy

The Government of Sri Lanka was requested by the International Union for Conservation of Nature and Natural Resources (IUCN) to take action to prepare a National Conservation Strategy for the country [12].

This strategy was developed by a high powered task force headed by the Chairman of the Central Environment Authority appointed by HE the President. This strategy itself was the outcome of 27 sector papers compiled by experts in all related fields and finalised

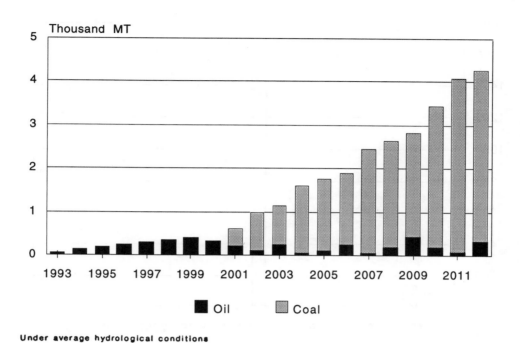

Figure 6.6. Power Sector Particulate emissions

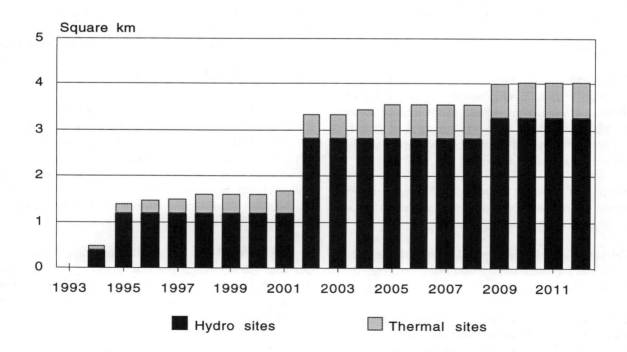

Figure 6.7. Land use

74

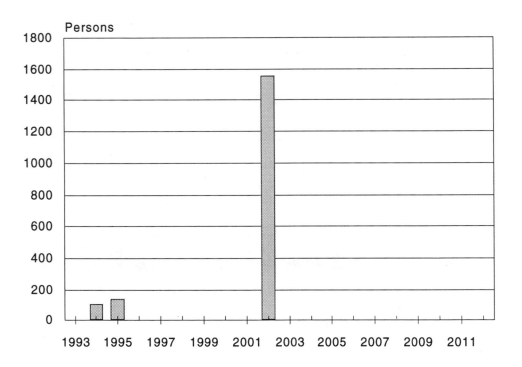

Figure 6.8. Resettlement requirements

after much deliberation and publicity. The extract which relates to energy sub sector is given in Annex A3.

8.0 Conclusions:

The paper clearly indicates that Sri Lanka will continue to grow as a developing country in the region and will require significant growth in the energy sector as well. Plans are drawn up to optimise the electricity supply system at least cost to the economy and with the least disturbance to the environment.

Supply of biomass is to be enhanced through forest conservation, afforestation and community wood lots, industry wood lots etc.

Petroleum supplies are geared to meet the demand and the refinery production is optimised to meet the product categories in the demand.

Demand side management measures are actively being pursued especially in the biomass usage through the National Fuelwood Conservation Programme.

Energy conservation in the industry and commercial sectors in particular is encouraged through the Ministry of Industries, Ministry of Energy Conservation and other relevant institutions. Energy

conservation in the transportation sector is pursued through legislation by restricting imports of over capacity cars and using fiscal measures such as lower tariffs for large size passenger transport vehicles which are more efficient in usage. These measures have led to conservation of petroleum in the country.

In the electricity sector transmission and distribution loss reduction programme and a programme to improve the load factor of the system along with proposed compact fluorescent light introduction programme will constitute some of the demand side management measures. The tariff structure too is adjusted to encourage conservation of electricity and shifting of the industrial load to off peak times.

Continued population growth which is between 1 per cent (1992) to 1.5 per cent (1991) [1] in the most recent past along with improvement of living standards of the people will continue to demand a growth in the energy sector.

The energy and environmental interaction is of prime importance in a country like Sri Lanka constrained by high population density and limited exploitation of hydro resources in the future. At the turn of century, the dwindling biomass resources, which is the main source of energy in the country, will have impacts on restrained fuelwood supplies in meeting the demand as well as having serious impacts on the environment.

The Government has introduced a national conservation strategy and an action plan through the Ministry of Environment. The Government has also enacted legislation through national environmental act of 1980 as amended in 1988 to safeguard against adverse environmental impacts due to development activities. Under this act regulations appointing project approving agencies for all development projects coming under various economic sub sectors are now in force.

Sri Lanka with this moderate economic and population growths will be able to sustain the much required energy resource inputs in the long term with minimum adverse environmental impacts.

References:

[1] Central Bank of Sri Lanka Annual Report 1992.

[2] Public Investment 1992 – 1996 Department of National Planning October 1992.

[3] Sri Lanka Peat Study, Ekono, Finnida, October 1995.

[4] K.K.Y.W. Perera, Energy Status of Sri Lanka Issues Policy Suggestions Policy Studies, March 1992.

[5] Sri Lanka National Report Ministry of Environment Parliamentary Affairs September 1991.

[6] R.R. Mel, National Fuel wood Conservation Project: Objectives, activities, evaluation and suggestions. Alternative Energy Development Branch, Ceylon Electricity Board, Sri Lanka, July 1993 (internal document).

[7] Sri Lanka Energy Balance and Energy Data, 1992. Alternative Energy Development Branch, Ceylon Electricity Board, Sri Lanka. (Yet to be published).

[8] Report on long term generation expansion planning studies 1994-2008, August 1993, Generation Planning Branch, CEB.

[9] Report on Masterplan for the Electricity Supply of Sri Lanka – June 1989.

[10] P. Meier and M. Munasinghe, Incorporative Environmental Concerns into Power Sector Decision – Making : A Case Study of Sri Lanka. Environmental Policy and Research Division, The World Bank, Washington DC, USA. June 1992.

[11] National Environmental Act No. 47 of 1980, Government of Sri Lanka

[12] National Conservation Strategy – Action Plan, Central Environmental Authorities, Ministry of Environment and Parlimentary Affairs, November 1990.

Impact of the replacement of 60 W incandescent lamps with 11 W CFLs

		IL	CFL		
Wattage (W)	=	60	11		
Life time (hrs)	=	1000	8000		
Cost (Rs.)	=	25	750		
No of lamps to be replaced	=	1,000,000			
Capacity saving /lamp (W)	=	(60 – 11)		=	49
Energy saving/lamp/yr (kWh)	=	49 x 4 x 365		=	72
Total capacity saving (MW)	=	49			
Total annual energy saving (GWh)	=	72			

Cost comparison

Exchange rate 1 US$	=	48 Rs.
Costs associated with GTs		
Capital cost	=	601.00 US$/kW
Fixed O&M cost	=	0.31 $/kW-mnth
Variable O&M cost	=	5.44 $/MWh
Fuel cost	=	2.13 Rs/kWh
Capital cost of 49 MW of GT	=	1413.55 M.Rs.
O&M cost saving	=	27.55 M.Rs./yr
Fuel cost saving	=	153.36 M.Rs./yr
Average selling price	=	3.30 Rs/kWh
Life time of CFL	=	5.5 yrs
Over the life time of CFL		
Saving on new GT	=	2408.56 M.Rs.
Revenue loss to CEB	=	1306.80 M.Rs.
Cost of CFL	=	750.00 M.Rs.
Net saving to CEB	=	351.76 M.Rs.

Incorporation of Environmental Issues into Power Generation Planning

Power sector long term planning is conventionally conducted at three levels: Generation, transmission and distribution. While minimising social and environmental impacts in the operation of the existing generating system, the future system offers the opportunity to fully consider the environmental and social objectives and to formulate mechanisms to minimise the impacts.

Different countries adopt different policies and methods to formulate future generation plans. Many developing countries adopt least cost planning techniques, in which the type, capacity and timing of new generating plants are selected to minimise the long term cost of electricity to the country. The objective is thus to minimise costs. Industrialised countries generally tend to plan future generation projects on a project-by-project basis, such that each project is selected on its technical and financial viability.

In the above processes, environmental and social impacts can only be included in the analysis in one way. That is by costing all the tangible impacts (such as loss of agricultural land in a hydroelectric reservoir) and costs of mitigation (such as the cost of flue gas desulphurisation in a coal plant) and by including them in the total cost function. Financially non-quantifiable impacts (such as CO_2 emissions) would therefore be only calculated and presented. Such impacts may then be justified by comparing with norms, emission standards or emission limits. If newly proposed measures such as Carbon taxes have been announced, such costs can be included in the total cost function.

Multi-attribute Decision Analysis

Multi-attribute methods have been used for some time in power plant siting and transmission line routing studies. Over the past decade this techniques has been used, for this purpose, in developing countries as well. At the same time, this method has been proposed for more general applications in energy and power sector planning.

A recent case study in Sri Lanka (Meier and Munasinghe, 1992)[10] has attempted to use the multi-attribute decision analysis techniques to the future expansion of electric power generation in the country. Sri Lanka uses the Wein Automatic System Planning (WASP) model for the planning and cost optimisation of the future power generating system. In the case study,

the multi-attribute decision technique has been developed to complement WASP results in the context of the environment.

Methodology

The first step is to select a series of attributes. The authors of the case study have chosen to limit the analysis to a few attributes, which are considered to be of higher significance. Concerns such as obstruction to scenic beauty by the presence of a coal power station or impacts on local households during transport of material to a hydro dam site have been left out to be treated during project-specific EIA studies

In the case of Sri Lanka, the following have been selected to be the 7 main environmental and social attributes.

(a) Emission of carbon dioxide

(b) Population exposure to air pollutants (SO_x and NO_x)

(c) Biodiversity

(d) Soil permeability

(e) Surface temperature (thermal plumes)

(f) Employment

(g) Emission of acid rain precursors

The case study also considers the usual exogenous assumptions in power sector planning. These are the demand forecast, system load profile, fuel price projections and the types/performance of the plants to be added to the system.

Figure A1 shows the modelling framework, showing the interaction of the multi-attribute analysis with the other power sector models. Figure A2 shows the schematic of the multi-attribute decision analysis software model- "Enviroplan". The software model conducts a capacity expansion plan and a merit order dispatch, thereby calculating the energy dispatch from each type of plant (both existing new) each year in the future. The environmental attributes are then calculated for each year in the planning period (20 years ahead).

There is a proliferation of policy measures and environmental/social safeguards that could be implemented, each of which would give rise to a new expansion plan and/or a new set of values for the 7 attributes selected. For example, in the current status of the generation system in Sri Lanka, if a policy of developing all remaining hydroelectric sites is implemented, the employment effect will increase, CO_2 emissions will decrease but the overall discounted system cost will increase.

The model then develops a scatter plot for each policy/environmental measure to show the cost of the resulting generation plan against the environmental attribute. Figure A3 shows such a plot of cost against CO_2 emissions. Each point in the plot refers to a different expansion plan, with the BASECASE as the reference case. From this result, it would be possible to develop a trade-off curve, which indicates the subset of most attractive expansion options from the objective of minimisation of cost and the environmental attribute. This process will then limit the analysis to a few policy options and their accompanying plans.

Each attribute will then result in a different trade-off curve, from which resource allocation decisions have to be made, depending on the relative importance of the attribute.

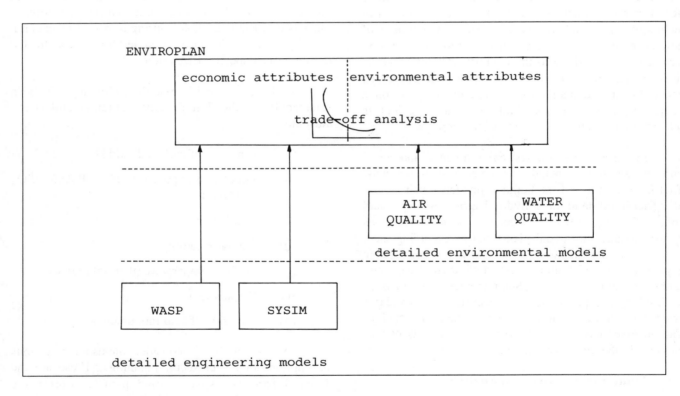

Source: Incorporation of Environmental Concerns into Power Sector Decision making [10]

Figure A1 : The modelling framework

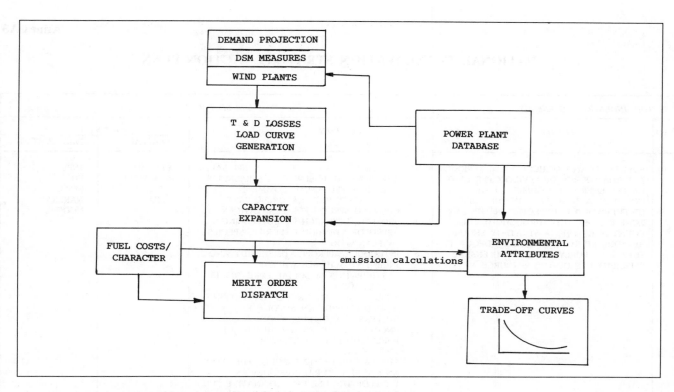

Source: Incorporation of Environmental Concerns into Power Sector Decision-making

Figure A2 : Major features of ENVIROPLAN

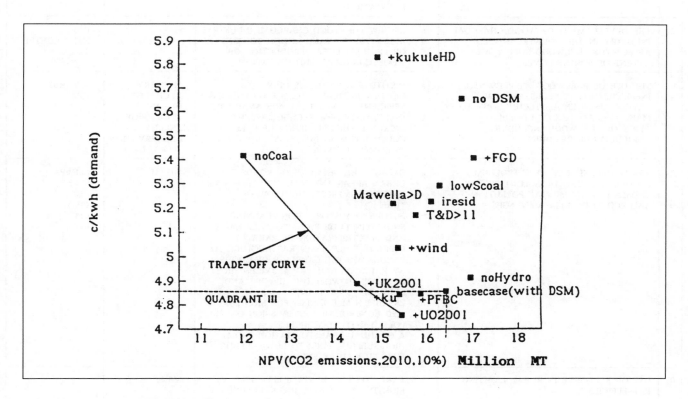

Source: Incorporation of Environmental Concerns into Power Sector Decision-making

Figure A3 : Greenhouse gas emissions

NATIONAL CONSERVATION STRATEGY – ACTION PLAN

SECTOR: INDUSTRY AND ENERGY		SUB SECTOR: ENERGY		
NO.	ISSUE	ACTION	AGENCY	
			PRINCIPAL	PARTICIPATING
01	FUELWOOD AND AGRICULTURAL RESIDUES ARE THE SOURCE OF SEVENTY PER CENT OF THE ENERGY CONSUMED IN THE COUNTRY. A SHORTFALL IN THE SUPPLY OF FUELWOOD IS EXPECTED IN THE DRY ZONE BY 1995 AND THROUGHOUT THE COUNTRY BY 2010. A STRATEGY SHOULD BE ADOPTED TO PROVIDE AN ADEQUATE SUPPLY OF FUELWOOD AND TO ENSURE EFFICIENT USE OF THE RESOURCE.	a. IDENTIFY APPROPRIATE AREAS, SPECIALLY AREAS WHERE FUELWOOD DEFICIENCIES ARE EXPECTED AND RECOMMEND APPROPRIATE HIGH CALORIFIC FAST GROWING SPECIES FOR PLANTATIONS. b. PROMOTE RESEARCH OF DEVELOPING APPOPORIATE HIGH CALORIFIC SPECIES OF PLANTATION; c.. INTRODUCE APPROPRIATE SUBSIDY AND INCENTIVE SCHEMES WHEREBY PARTICIPATION IN SOCIAL FORESTRY IS ENCOURAGED; d. ENSURE THAT FUELWOOD PLANTATION WOULD PROVIDE AN ADEQUATE BUFFERSTOCK OF FUELWOOD; e.. ENCOURAGE THE ESTABLISHMENT OF FUELWOOD LOTS, WHEREVER APPROPRIATE; f. REVIEW THE PRESENT USE OF FUELWOOD AND AGRICULTURAL RESIDUES AND DEVELOP THE USE OF ALTERNATIVE FUEL SUCH AS COIR DUST, BAGASSE AND PADDY HUSKS EVOLVING OPTIMUM USE STRATEGY; g. ENSURE THAT WHEREVER POSSIBLE INDUSTRIES USING FUELWOOD AS FUEL HAVE THEIR OWN FUELWOOD PLANTATIONS.	M/LI&MD MAHAWELI AREAS – M/MD	M/PI FD MASL NARESA NGOO
02	THE DEVELOPMENT OF WOOD CHARCOAL INDUSTRY IN THE COUNTRY WOULD AFFECT THE FUELWOOD SUPPLY AND EXTENT OF FOREST COVER.	a. REVIEW THE WOOD CHARCOAL INDUSTRY; b. IF NECESSARY, TAKE ACTION TO CONTROL OR PROHIBIT LOCAL PRODUCTION OF WOOD CHARCOAL AND ITS EXPORT.	FD	IDB NERD CENTRE TC
03	THE USE OF FUELWOOD AT PRESENT IS INEFFICIENT. IT IS NECESSARY TO DEVELOP INDIGENOUS TECHNOLOGY TO MAKE FULLEST USE OF DIFFERENT TYPES OF FUELWOOD INCLUDING AGRICULTURAL RESIDUES.	a. INSTITUTE A VIGOROUS DRIVE TO RAPIDLY INCREASE THE USE OF ENERGY-EFFICIENT STOVES THAT ARE AVAILABLE; b. INVESTIGATE NEW TECHNOLOGY TO INCREASE THE EFFICIENCY OF FUEL CONSUMPTION IN INDUSTRY AS WELL AS IN DOMESTIC AREA.	MINISTRY IN-CHARGE OF THE SUBJECT OF ENERGY CONSERVATION	M/T&RID IDB
04	HEAVY DEPENDENCE OF PETROLEUM PRODUCTS IN THE TRANSPORT SECTOR CAUSES ENVIRONMENTAL POLLUTION AND ECONOMIC DISADVANTAGES.	a. EXPAND THE NETWORK OF ROADS IN URBAN AREAS AND MAINTAIN EFFECTIVE LANES TO IMPROVE THE FLOW OF TRAFFIC; b. FUTURE TRANSPORTATION PLANNING SHOULD INCLUDE RAIL – ROAD LINKAGE AND ELECTRIFICATION OF RAILWAYS; c. IMPROVE RAILWAY INFRASTRUCTURE TO ENCOURAGE THE CARRIAGE OF GOODS BY RAILWAY. RAILWAY SERVICES SHOULD BE PLANNED WITH THE AIRPORT AND PORT AS NODAL POINTS; d. ESTABLISH MECHANISM FOR MONITORING AND CONTROL OF THE EMISSION OF POLLUTANTS BY VEHICLES; e. A GRADUAL TRANSFORMATION FROM HIGH-SULPHUR FUEL TO LOW-SULPHUR FUEL SHOULD BE ENCOURAGED.	M/T&H	M/E&PA M/P&E CGR C/MT CPC RDA
05	USE OF KEROSENE IN RURAL LIGHTING IS INEFFICIENT.	a. TAKE STEPS TO DEVELOP EFFICIENT LIGHTING SYSTEMS USING KEROSENE.	M/P&E	

NATIONAL CONSERVATION STRATEGY – ACTION PLAN (continued)

SECTOR: INDUSTRY AND ENERGY		SUB SECTOR: ENERGY		
NO.	ISSUE	ACTION	AGENCY	
			PRINCIPAL	PARTICIPATING
06	ADDITIONAL HYDRO-POWER AS A SOURCE OF POWER GENERATION MAY NOT BE ECONOMICALLY VIABLE IN THE FUTURE. THE USE OF OIL IN POWER GENERATION IS LIMITED BY SPIRALLING COSTS. THEREFORE, THE INCREASING POWER DEMAND WILL HAVE TO BE MET BY ALTERNATIVE POWER GENERATION. ALTERNATIVE SOURCES OF POWER GENERATION SUCH AS COAL, DENRO, AGRICULTURAL RESIDUES, ETC., NEED TO BE CONSIDERED.	a. REVIEW AND UPDATE THE POWER REQUIREMENTS OF THE COUNTRY; b. IDENTIFY NEW SOURCES OF POWER GENERATION WITH RESPECT TO A TIME FRAME; c. EXPAND THE NETWORK OF POWER DISTRIBUTION WITH RESPECT TO A TIME FRAME; d. ASSESS THE ENVIRONMENTAL IMPACTS LIKELY TO ARISE FROM THE IMPLEMENTATION OF POWER AUGMENTATION SCHEMES AND DRAW UP ACTION PROGRAMMES TO MITIGATE THE ADVERSE IMPACTS; e. REDUCTION OF LOSSES IN TRANSMISSION AND DISTRIBUTION OF POWER SHOULD BE IMPLEMENTED ON A PRIORITY BASIS.	CEB	M/E&PA MINISTRY IN-CHARGE OF THE SUBJECT OF ENERGY CONSERVATION M/P&E
07	NON-CONVENTIONAL SOURCES OF ENERGY SUCH AS SOLAR ENERGY, WIND ENERGY AND BIOGAS HAVE NOT SO FAR BEEN WIDELY ACCEPTED. A STRATEGY SHOULD BE ADOPTED TO MAKE THEM VIABLE SOURCES OF ENERGY.	a. ENHANCE RESEARCH AND DEVELOPMENT EFFORTS IN RESPECT OF THESE POTENTIAL SOURCES OF ENERGY GENERATION; b. ESTABLISH AN INSTITUTIONAL FRAMEWORK FOR TRAINING, TECHNOLOGY TRANSFER AND DISSEMINATION OF INFORMATION.	MINISTRY IN-CHARGE OF THE SUBJECT OF ENERGY CONSERVATION	M/P&E

CHAPTER VII

PROSPECTS FOR ENERGY CONSERVATION AND EFFICIENCY
IN THE ESCAP REGION[*]

A. Energy efficiency : precondition for sustainable development.

The availability and utilization of energy is as much a prerequisite as it is an integral part of economic expansion modernization and growth. The developing countries in particular, including those of the ESCAP region, will continue to require more and additional energy resources, mainly fossil fuels, as they seek to achieve higher levels of industrialization and productivity, accelerated growth of gross national product and increased output, and to improve national and social welfare. Throughout the ESCAP region, Governments, energy planners and utilities therefore expect significantly increasing levels of energy demand and consumption.

However, emissions resulting from the combustion of fossil fuels are a major source of environmental pollution of land, water and air land. On a local scale, examples are photochemical smog, referring to the mixture of unhealthy gases formed when solar light acts on gaseous emissions from combustion engines, effluent discharge into water courses and waste disposal on land. On a regional scale, impacts include acid precipitation, from emissions such as SO_2 and NO_x, leading to depreadation of plants growth, soil acidification and acidification of surface waters.

Several emissions combine to raise global temperature, an effect known as 'greenhouse warming'. Greenhouse gases (mainly carbon dioxide CO_2, methane CH_4, nitrous oxide N_2O and chlorofluorocarbons) absorb infrared radiation that becomes trapped as heat in the atmosphere. In the past decades the concentration of greenhouse gases in the atmosphere has increased tremendously, on the longer term leading to an increase in average global surface temperature, and accompanied by effects such as sea-level rising and changes in precipitation patterns. To stabilize atmospheric concentration of greenhouse gases at today's level would require immediate reduction.

The accelerating economic expansion process which is presently taking place in many developing countries, particularly in Asia and its large and growing metropolitan urban and industrial centres, requires large amounts of additional energy resources, mainly fossil fuels, and will cause considerable environmental impacts. It has been estimated that Asian and Pacific developing countries would annually add 2.5 gigatonnes (Gt) of carbon equivalent by the 2010, three times the current rate. With maximum feasible intervention (including energy efficiency improvements and fuel switching) this could be reduced to no less than 1.8 Gt per year[1].

The largest source of greenhouse gas emissions is combustion of fossil fuels, and therefore any emission reduction strategy should aim at least cost energy approaches, fuel switching, improved energy conversion efficiencies and improved energy end-use efficiencies. Especially, improvements in energy end-use efficiency are generally regarded as being both technically feasible and economically viable, often having net economic benefits (i.e. savings) and relatively short pay-back periods.

Apart from environmental reasons, Asian countries can take advantage of enhanced energy economy and efficiency in terms of reduced investments in energy supply infrastructure, reduced dependence on energy imports, reduced strains on the balance of payments, delayed depletion of scarce energy reserves and stimulation of indigenous technical competence and industrial capacity. Enhanced intermediate and end-use energy efficiency in the industrial, service and transport sectors of developing and developed countries can strengthen national and international competitiveness, facilitate economic expansion and enhance diversification through new and additional business opportunities.

This chapter presents a comparative introductory overview and a summary assessment of the varying energy intensities in the countries of the region. As part

[*] This chapter is a revision of the secretariat note (NR/PCSED/5) presented at the Expert Group Meeting Preparatory to the First Session of the Committee on Environment and Sustainable Development, Bangkok, 30 September-2 October 1993.

[1] *Energy Policy Implications of the Climatic Effects of Fossil Fuel Use in the Asia-Pacific Regions,* ESCAP, United Nations, November 1991, p.5.

of a brief sectoral review, the most energy intensive activities and industries are identified. Energy conservation potential is discussed. The chapter looks at the main barriers to energy efficiency improvements, and it outlines essential elements for policies and programmes.

B. Energy intensity : synopsis of historic and structural trends

1. Regional overview

The data presented in table 1 and figure I are meant to give an introductory overview of the prevailing average energy intensities, economic activity, population and energy consumption growth. The review focuses on the larger developing economies of the region, where energy requirements and energy consumption are growing fast, and where, therefore, conservation potential might be more significant. For purposes of an easy comparison, relevant data are also presented for selected industrialized countries of the Pacific, France and the United States of America.

As compared to the industrialized countries, where the average annual energy consumption per capita generally ranges between 1.5 and 3.5 tonnes of oil equivalent/capita/annum,[2] consumption in the developing countries and areas of the ESCAP region is still comparatively low, with the exception of Hong Kong, the Republic of Korea, Singapore and Taiwan Province of China (not included in the table), where industrialization, consumption patterns and lifestyles are similar to those of industrialized countries.

Figure I illustrates energy intensities in GDP (in constant 1980 United States dollars) across selected regional countries, showing a slight upward trend in the past 20 years (with the exception of China).

Recent trends in the development of energy intensity between 1980 and 1990 are reflected in figures II (a), (b) and (c) in relative terms, to remove the influence of foreign exchange fluctuations. Figure II groups data on national energy intensity trends (commercial energy consumption permit of GDP, in 1980 domestic currency): (a) For ASEAN (Association of South-East Asian Nations); (b) For other Asian countries; and (c) For selected industrialized countries.

(a)   During the period from 1980 to 1990, average energy intensities of economic activities in most

ASEAN countries and large, remained unchanged. The fall in oil prices resulted in decreasing interest in energy conservation. By the end of the decade, energy intensities seem to have increased in several countries as a consequence of rapid economic and industrial growth. In countries where industrialization has commenced earlier and/or has grown further, as well as in many of the newly industrializing economies (Hong Kong, Republic of Korea, Singapore) energy intensities have also gradually decreased and the gradually increased towards the end of the decade. In these economies, the relatively less energy-intensive service sectors also increased their contribution to the total national value added. Modernization seemingly has made manufacturing processes more productive and energy efficient.

(b)   As compared with the ASEAN countries, the development of the average energy intensity in many of the other Asian countries show similar trends, except for Bangladesh and the Islamic Republic of Iran. In Bangladesh, the emerging structural change and the growing role of the non-agricultural sectors may have been important reasons for the gradual increase in energy intensity. In the case of the Islamic Republic of Iran the increasing energy intensity towards the end of the decade may reflect the resumption of economic growth after the war between Iraq and the Islamic Republic of Iran. The increase in energy intensity in the Islamic Republic of Iran in 1989 may be regarded as a normalization of economic development after the war. In China, the energy use per unit of output value has remained high despite the energy conservation measures implemented and the efficiency gains achieved by some of the industrial sectors[3], but energy intensity is gradually decreasing.

(c)   To a considerable extent, energy intensity appears to be a function of modernization and economic development. The varying energy intensities also reflect the varying resource endowment and varying energy policies. The industrialized and the newly industrializing countries of the Asian and Pacific region and the overall structural pattern of their industrial and economic development may suggest, prima facie, that the developing economies, too, will be able to curb the increases in energy intensity in the more distant future, as higher levels of productivity in industrial manufacturing are achieved, more value added is produced in the service sectors, and more and improved

---

[2]   Per capita consumption of energy is particularly high in the United States of America, and also in Australia.

[3]   However, large parts of the energy-intensive industries in China have remained under direct government administration. For some of the industrial sectors, energy is still directly or indirectly subsidized; hence, the high average energy intensity.

Table 1. Indicators of economic growth, population, energy consumption and energy intensity: selected industrialized and developing countries and areas (1990)

	GNP (in billion 1990 constant US$)	Energy consumption (in million metric toe)	Population (in million)	GNP/capita (in 1990 constant US$)	Energy consumption/capita (kg of oil equivalent)	Energy intensity (kg oil equivalent/$ of GNP)
Asia						
Bangladesh	22.8	5.9	106.0	214	51	0.238
China	3 64.0	645.5	1 134.0	321	569	1.773
Hong Kong	70.1	7.2	5.8	12 086	1 240	0.103
India	280.4	185.4	850.0	330	217	0.658
Indonesia	102.3	38.8	178.0	574	211	0.367
Iran (Islamic Republic of)	118.8	63.2	55.8	2 128	1 158	0.544
Malaysia	40.0	18.9	17.9	2 268	1 055	0.465
Pakistan	38.7	23.8	112.4	353	194	0.548
Philippines	43.8	13.1	61.5	714	210	0.284
Republic of Korea	236.8	74.1	42.8	5 535	1 731	0.313
Singapore	35.3	10.6	3.0	11 768	3 878	0.330
Sri Lanka	8.0	1.5	17.0	470	91	0.184
Thailand	78.3	28.7	55.8	1 422	516	0.363
Pacific						
Japan	2 864.1	358.5	124.0	23 804	2 904	0.121
Australia	283.0	88.9	17.1	16 550	5 271	0.318
New Zealand	40.2	11.9	3.4	11 812	3 497	0.286
Other countries						
United States	5 441.0	1 737.2	250.0	21 764	6 970	0.320
France	1 186.7	155.9	56.4	21 040	2 776	0.132

Source : United Nations, 1990 Energy Statistics Yearbook, New York, 1992. World Bank, World tables, 1992, Washington, 1993.

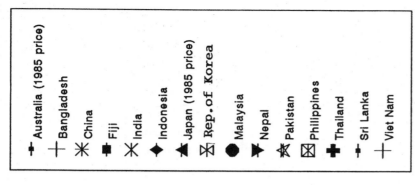

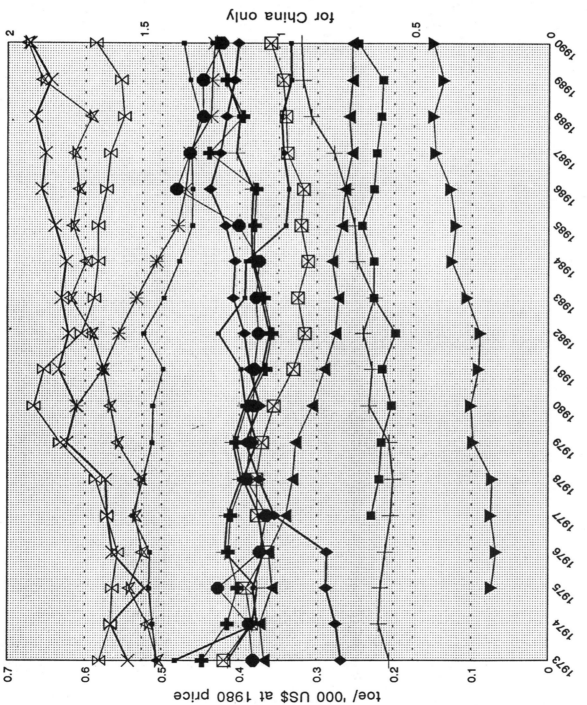

Source: ADB Energy Indicators 1990, IEA Statistics for Australia and Japan.

Figure 1. Energy Intensity in Gross Domestic Product

86

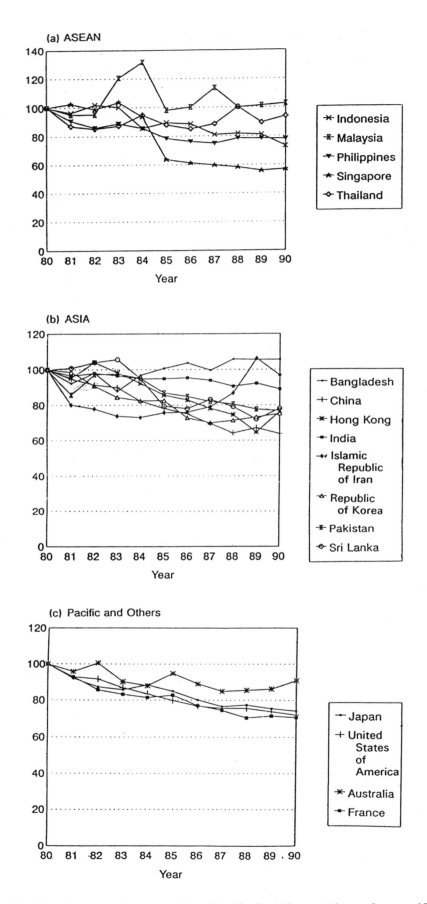

Figure II. Development of energy intensity of selected countries and areas 1980-1990

infrastructures are established. However, the economic development in the industrialized and in the industrializing countries has shown that economic success is not a guarantee for improved energy efficiency.

2. Sectoral overview

(a) Industrial manufacturing

At the global level, the manufacturing sector accounts for the largest share of final energy use. Depending on the degree of industrialization and on the structural composition of the sector, manufacturing accounts for between 25 and 45 per cent of final energy use. In the developing countries, the relative share of the manufacturing sector in energy use in 1990 varied between 22.6 per cent in Sri Lanka, 28.5 per cent in Thailand, 33.5 per cent in the Philippines, 37.9 per cent in the Republic of Korea, 39.9 per cent in Malaysia, 41.9 per cent in India, 47.3 per cent in Indonesia, and 66.7 per cent in China.

Among the various groups of industry, production of ferrous metals is one of the most energy-intensive lines of production. The pulp and paper industry, manufacturing of non-ferrous metals, building materials such as cement, and chemical industries are also more energy-intensive than other industrial sectors. In the developing countries some of the small and medium-scale industries are also energy-intensive (for example, ceramics factories, mini-foundries, manufacture of bricks, some forms of food-processing).

Over the past 20 years, technical progress has led to increasing productivity, a process that has often been accompanied by gains in energy efficiency. Figures III-A and III-B show the energy intensity of selected industrial sectors for Japan and China.

Industrial manufacturing in industrialized countries is not necessarily more energy-efficient than manufacturing in developing countries: some of the more rapidly industrializing countries of the ESCAP region have, in fact, developed modern industrial complexes with energy-efficient technologies. However, the prevailing capital shortage has prevented a number of other countries from expanding and modernizing their industrial plants.

In the developing countries of the ESCAP region, small and medium-scale industries also play an important role, not only in employment generation, but also in energy consumption. It is in particular the small and medium-scale industries sector where (owner-)

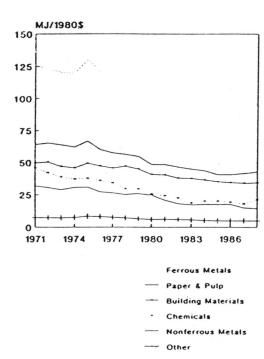

Source: Lee Schipper, Stephen Meyers and others, *Energy Efficiency and Human Activity: Past Trends, Future Prospects,* (1992), p. 96.

Figure III-A Japan: Energy intensity by group of industry

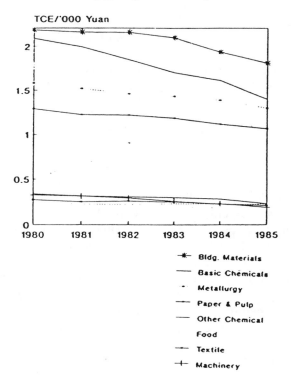

Source: Lee Schipper, Stephen Meyers and others, *Energy Efficiency and Human Activity: Past Trends, Future Prospects,* (1992), p. 105.

Figure III-B China: Energy intensity by industry subsector

managers, due to the capital constraints, tend to choose the less and least expensive equipment. The medium and longer term implications of the prevailing preference for the less capital intensive (and low energy-efficient) technologies, equipments and products are that the developing economies face a critical and growing risk of structuring their investments and their entire technological systems in ways which may make them become (re)producers of long-term low energy-efficient high energy consumption techniques.

(b)   Transport

During the past two decades per capita travel increased considerably throughout the world and in all countries of the ESCAP region. As in other parts of the world, there has been a shift in the mode of transport towards automobiles, and in some countries towards airplanes, as well. This shift has caused growth in the aggregate energy intensity of travel.

The structure of travel in most developing countries shows a similar pattern, with most people relying on bus, rail and non-motorized modes of transport, while the wealthier sections use cars. Buses and collective taxis are important means of transport. Modern energy-efficient mass transit systems are still lacking in most Asian metropolitan cities. In much of Asia, there has been growing use of mopeds and motor/cycles, and thus, more inequity in transport in other countries compared with industrialized countries. This can be addressed only by discouraging the use of cars and promoting mass transit systems.

For the different countries of the region the development of automobile energy intensity has been difficult to judge. In many cities, the population of vehicles, buses, motor/cycles and automobiles has grown tremendously. Newly produced vehicles tend to be more energy-efficient, thus reducing the average energy intensity of passenger travel. However, road vehicles are often kept in service for longer than they were designed for (due to capital scarcity) and these ageing vehicles are often poorly maintained (because of spare parts scarcity). At the same time, the often inadequate infrastructure (poor roads in rural areas and traffic congestion, the expanding urban-metropolitan areas) adds to a high energy use per person kilometre in automobile travel..

Like passenger transport, freight activities have also grown significantly throughout the ESCAP region. The modal structure of freight transport in the region varies across countries, and changes with the composition of economic output and national geography.

In the less and least developed countries animal-powered transport of goods is still important. In other countries, like Indonesia and the Philippines, many goods are transported by ships.

China and India have extensive rail networks. The energy intensity of rail transport has declined in India due to increased use of diesel and electric locomotives, which have partly replaced the inefficient coal-fired steam locomotives.[4] In China, steam trains are still predominant, though the growing use of diesel locomotives has improved energy efficiency.

Throughout the region, there has been a shift from rail to trucks. Transport by trucks is more energy-intensive, but provides greater flexibility and other advantages for moving manufactured products. In some countries, there has been a shift from gasoline to diesel trucks, which may have contributed to some reduction in energy intensity.

Thus, the growth and the modal changes of the transport sector, except perhaps for China and India, have considerably increased oil consumption and oil import dependency. There is a need to promote rail transport. However, the difficulty lies in the fact that rail transport is normally managed by the public sector, as opposed to land transport, which is a private sector activity. Private sector activities in land transport expand more easily.

(c)   Service sectors

The service sector, also often referred to as the commercial sector, comprises activities such as wholesale and retail trade, finance, insurance and real estate, business and personal services, restaurants, hotels and hospitals. Most energy use in this sector takes place in buildings of one kind or another.

In the developing countries of the ESCAP region, the share of the service sector in final energy use has grown in recent years. Relative to other sectors, the service sector accounts for a relatively large share of electricity consumption. In Japan and in the newly industrialized and industrializing countries of the region, the service sector and its electricity use is driving peak demand, in particularly during the hot summer months.

In the countries of the region a substantial fraction of energy and electricity use in the service sector goes for air-conditioning (and/or space heating) of buildings,

----

[4]   See Lee Schipper and Stephen Meyers, and others, *Energy Efficiency and Human Activity: Past Trends, Future Prospects* (Cambridge, 1992).

including office buildings, schools, shopping complexes, or hospitals. In the warm and humid climates characteristic of many of the developing countries in South-East Asia, air-conditioning is a major consumer of electricity. Many of the modern buildings are more energy-intensive than older buildings. Recent studies have shown that electricity intensity in kilowatt hour per United States dollar of GDP has grown in many of the ESCAP developing countries.[5]

In the developing countries of the region the continuing expansion of electricity generation capacities is becoming increasingly costly. Hence,improvements in energy conservation and efficient building energy management are important areas for achieving higher energy efficiency.

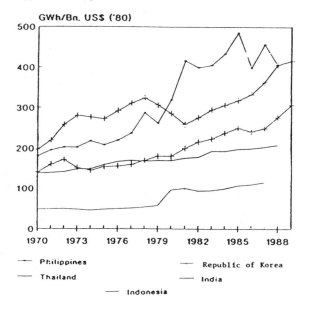

Figure IV. Service sector electricity intensity – selected countries of the ESCAP region

(d)   Residential energy end use

In the developing countries, the residential sector accounts for approximately 15 per cent of energy end use.[6] The nature of residential energy use, the historic trends, the relative share of the fuels used and the degree to which the desires for new services and appliances are met differs significantly from country to country. Whereas most households in the industrialized countries already acquired many amenities, most households in the developing countries have not yet been able to upgrade

their living and/or comfort standards. Indeed, many are not yet connected to the grid. The energy demand of the household sector in the developing countries of the region will grow considerably during the years to come.

The past, present and future energy demand of the residential sector is mainly determined the degree of overall economic growth and the growth in the (real) disposable income. In addition, the changing lifestyles are also found to influence the patterns of energy demand and energy use: as in the industrialized countries, in some developing countries of the region the average number of persons per household starts to show a gradual decline. Whilst larger households are able to utilize economies of scale for certain energy end use, appliances in smaller households tend to require similar amounts of energy as in large households. Hence, per capita energy use is found to be higher in small households.

In the lower-income countries, and among the poor in most developing countries, cooking accounts for the major share of residential energy use. With increasing urbanization and income households in developing countries tend to switch from traditional biomass fuels to kerosene and to liquefied petroleum gas (LPG),and, in some cases, electricity for cooking. Kerosene and LPG stoves are more energy efficient than biomass cooking.

In most of the developing country households, the main use of electricity is for lighting and electric household applian us. In the urban areas, lighting energy use has been rising along with the average number of lamps per home. Apart from lighting, fans, television sets and refrigerators are widely used electric household appliances. In some of the developing countries of the region more than half of all households are already equipped with television sets and refrigerators. Most developing countries neither have nor enforce energy efficiency standards for household appliances. Whereas new models of electric appliances sometimes tend to be more energy-efficient than older models, they also often tend to have more features, which can negatively affect energy efficiency.

In a recent study on Indonesia, it was found that new models of electrical appliances are sometimes still well behind the state-of-the-art in the manufacturers' home country.[7] The trade in non-energy-efficient technologies and products therefore appears to be a critical issue which may need to be addressed by energy policy makers.

---

[5]   Schipper, Meyers, and others, op. cit.

[6]   Not including biomass fuels. If biomass fuels are included in the calculation, the share of the residential sector in final energy use may vary between 20 and 40 per cent of total energy use.

[7]   Lee Schipper and Stephen Meyers, *Improving appliance efficiency in Indonesia,* Energy Policy, 19(6), (1991), pp.578-588.

In the industrialized countries households are also equipped with many other appliances, including air-conditioners and/or space heaters, water heaters, washing machines and dishwashers, freezers, as well as a large variety of other home office equipment, music players and other entertainment appliances. Many of these appliances are now also produced in the developing countries. However, more often than not quality and energy efficiency are found to be lower than in counterpart products produced in or for the markets of industrialized countries.

C. Potential for energy conservation and energy efficiency gains

For a realistic assessment of the potential for energy conservation and energy efficiency gains it is essential to distinguish what is technically or theoretically possible from what is economically desirable.[8] Many theoretical options may exist to increase energy efficiency. In overview of possible energy end-use conservation options is given in table 2. However, the number of options which consumers may be willing to consider is more limited. Here again, one needs to distinguish between what is socially desirable (taking external costs and benefits into account) and what is economically beneficial from the individual producer's or consumer's point of view.

Whereas some energy efficiency gains will be attained along with technical progress, there will also always remain a differential between socially desirable and individually achieved efficiency levels. However, the width of the gap depends on government policies and on their control and guidance of individual investment and consumption decisions.[9]

1. Industrial manufacturing

The experience has shown that technical progress is often associated with naturally occurring energy efficiency gains. In the developing countries manafacturing energy intensities can therefore also be expected to decline over time as existing facilities are improved and new, modern plants with higher

productivity and energy efficiency are built. Technically proven energy conservation techniques, as well as recycling and resource efficiency, could save an estimated 10-30 per cent of industrial energy[(*)]. On the other hand, many developing countries will expand energy and material-intensive industries as well as their infrastructure. This modernization process will vary from country to country and be influenced by a variety of factors, including (a) competition and openness of markets, which tends to accelerate technological adaptation and progress; and (b) economic growth and industrial integration which can bring economies of scale in production and lead to decreasing energy intensities. Energy pricing, industrial development policies and the introduction or absence of incentives for reduced energy consumption will also be important factors which determine the prospects for energy efficiency.

In most countries of the ESCAP region many energy audits have been undertaken, either for individual plants, or for certain industrial processes. These studies have shown that there is considerable potential for cost-effective improvements in energy management and/or retrofit investments.

The industrial sector is most varied in terms of energy end-use, products produces, processes employed, as well as size, age and facility conditions, industries range from handicrafts (e.g. small-scale metal working, carpentry, weaving, brick making) to light industries (e.g. shoes, textiles, pulp and paper, food processing, glass, brick making), heavy and energy-intensive industries (e.g. processing of metals, cement), mining and feedstock industries (e.g. chemicals, fertilizer). Fuel use is also varied (wood, bagasse, coal, diesel and gas oil, fuel oil, natural gas, electricity). On the other hand, the number of end-uses to which energy can be divided is actually quite small. Main end-uses include electric motors, electricity for electrolysis (e.g. aluminium) and direct consumption of fossil fuels for heating and drying in materials processing.

Ways to increase the economic efficiency of energy use are energy conservation and substitution. Revisions of operating and maintenance procedures (e.g. shutting off stand-by furnaces, regular calibration of instruments and regular inspection and repairing) are examples of short-term measures that are almost always extremely cost-effective. Longer-term measures include changes in processes that require plant redisign and thus

---

[8]   See also Fereidoon Sioshansi, *The myths and facts of energy efficiency,* Energy Policy, 19(3), 1991, pp. 231-243.

[9]   The International Energy Agency has recently published a comprehensive assessment of the energy efficiency potentials in the Organisation for Economic Cooperation and Development (OECD) countries. A summary of this assessment is attached to this paper in the annex. At this stage, no similar comprehensive assessment of potential seems to be available for the developing countries (on non-OECD countries).

[(*)] Footnote : "Greenhouse for Emissions and the Developing Countries : Strategic Options and the USAID Response", Report to Congress, USAID, Washington DC, 1990.

larger investments). Longer-term conservation measures are:

1) Recovery and reuse of waste heat. High-temperature heat (650°C) is produced in furnaces (steel making, copper refining, glass melting), incinerators and cement kilns. Medium-temperature heat sources (230-650°C) are turbine and engine exhausts and flue gases from boilers or ovens. Low-temperature heat sources are condensate and cooling water from high-temperature processes. high- and medium-temperature heat can be used to produce steam to drive a turbogenerator or supply process steam for pre-heating.

2) Improvements in the electric system. Apart from lighting, most electric equipment is in the form of motor power drives (such as pumps, fans, blowers, compressors and various tools) and electric furnaces and electrolysis. Electricity conservation can be achieved by better lighting techniques, reduction of a factory's peak load (by rescheduling operations, automatization and replacing oversized or inefficient equipment) and introduction of variable speed drives (VSDs). VSDs are electronic devices that enable a motor to vary its speed in order to match the speed of the motor to power demand.

3) Shifts to more efficient processes and materials. Opportunities exist in retrofitting of current designs where this is possible and feasible. Further, developing countries could adopt the most efficient processes or even currently experimental technologies ("technological leapfrogging"). Finally, large savings are available from the increased use of recycled materials.

4) Cogeneration. This is the combined production of electricity and steam (for on-site heating requirements). Current practice is that heat and electricity are often produced separately.

In a conventional thermal plant about 30-40 per cent of the energy content of the fossil fuel used is converted into electricity and the rest is discharged into the environment, as it is of too low temperature to generate electricity economically. In industries heat is usually provided by direct combustion of fossil fuels. An alternative for industries is to use the fossil fuel to generate electricity while the waste (low-grade) heat is used for process requirements. In this way overall efficiency of the fuel can be increased to 60 per cent. The additional advantage is that if all economic generation possibilities of a country are used, then a large fraction of a nation's demand can be supplied by decentralized cogeneration facilities instead of by the main grid Cogeneration could produce more than a 10 per cent increase in electricity-generation capacity in developing countries [Sharafi (1987)].[*]

In the energy-intensive large-scale manufacturing processes potential energy efficiency gains are also substantial. However, the potential for conservation varies from sector to sector and from project to project: in steel making and in non-ferrous metals, significant energy savings can be made by integrating production processes such as continuous casting, integrated rolling and finishing. In the pulp and paper industry, mechanical and chemical processes require electricity and heat in large quantities. In pulp and paper production, considerable amounts of energy can be save by modernizing equipment and by recycling waste paper. In the production of cement, energy accounts for approximately half of the total manufacturing cost. Modern cement plants using the dry or semi-dry clinker manufacturing process tend to require less energy than plants based on the wet process. Established manufacturing processes cannot be change or modified easily. The eventual introduction of the above mentioned processes may therefore be regarded as medium- and long-term options.

2. Transport

With the continuing changes in lifestyles, growing division of labour, greater incomes, professional specialization and greater product diversification, the transport sector will continue to grow considerably, in particular in the developing countries, both in the areas of passenger travel, as well as for the transport of goods.

Almost all car manufacturers have developed high fuel-efficient prototype vehicles. Fuel efficiency can be expected to increase in new cars. However, prospects for actually achieving energy efficiency will continue to depend mainly on the (expected) development of real energy prices, in particular for oil. In this context, prospects for attaining higher fuel and energy efficiency

---

(*) Footnote : Sharafi, "Cogeneration in Developing Countries : Prospects and Problems," in : Natural Resources Forum 11(1), UN, New York, USA, 1987.

Table 2. List of possible energy end-use conservation options

SHORT-TERM OPTIONS

Industry sector

- promotion of further efficiency improvements in production processes
- materials recycling (in particular energy-intensive materials)
- substitution with lower energy-intensive materials
- improved electromechanical drives and motors
- thermal processes optimization, including energy cascading and cogeneration
- improved operation and maintenance

Energy storage

- Improved cogeneration

Transport sector

- improved fuel efficiency of road vehicles
  - advanced vehicle design
  - electronic engine management
  - regular vehicle maintenance
  - improved efficiency transport facilities
  - regenerating units
- technology development in public transportation
  - intra-city shift to public transport
  - intercity trains
  - better intermodal integration
- imporved traffic management, driver behaviour, vehicle maintenance

Housing and building sector

- improved heating and cooling systems
  - improvement of energy efficiency of air conditioning
  - promotion of area heating and cooling, include the use of heat pumps
  - improved burner efficiency
  - use of heat pumps in buildings
  - use of control systems
- improved space conditioning
  - improved heat efficiency through highly efficient insulating materials
  - better design (orientation of windows, building envelope, etc)
- improved lighting efficiency
- improved efficiency of appliances
- improved operation and maintenance

MEDIUM-TERM OPTIONS

Industry sector
- increased use of energy-efficient materials
- advanced process technologies
- use of biological phenomena in processes
- localized energy conversion

Transport sector
- improved fuel efficiency road vehicles

Energy storage

- advanced batteries

Housing and building sector
- improved energy storage systems
  - use of informaiton technology to anticipate and satisfy energy needs
- Improved building systems
  - new building materials
  - windows that adjust opacity

POLICY OPTIONS

Technology transfer

- information exchange (through conferences, workshops, clearinghouses, training and education programmes, customer informaiton, extension services,)
- improved energy services (e.g. private sector involvement)
- energy efficiency standards, labeling

Economic/financial measures

- appropriate energy pricing
- subsidies
- taxes and levies
- adjust import restrictions on energy-efficient technology

in the transport of passengers and goods in the future will depend to a considerable extent on decisions of national governments today: energy and oil pricing, taxation, speed limits and many other policy instruments are available to "govern" the development of the transport sector.

Congestion on the one hand acts as a deterrent by discouraging the use of cars, and forces people to use mopeds/cycles, but on the other hand increases fuel use per km of vehicles and has economic consequences [time (lost) is money (lost)].

More extensive roads alleviate congestion, but from an energy conservation as well as transport efficiency point of view should be combined with offering alternatives, such as mass transit systems (subway, intra-urban railway) and efficient bus systems. The cities of Singapore and Hong Kong have shown how urban mass transportation systems can be established and maintained successfully.

3. Services sector

The services sector will also continue to grow throughout the region. In the modern commercial buildings the use of air-conditioning and office equipment will continue to increase. In the developing countries with tropical climates the cooling load in modern-style buildings is often very high. In the services sector, the areas with the noteworthy potential for energy conservation and efficiency improvement are lighting and air-conditioning. The installation of energy efficient lighting (high efficiency fixtures, ballasts, and lamps) as well as improved building techniques, that take advantage of day-lighting, can reduce not only direct electricity consumption, but also overall cooling loads, which may allow downsizing of HVAC (heat, ventilation and air-conditioning) equipment.

In the services sector, there are also high energy savings possible for air-conditioning through the application of other commercially available technologies, such as high-efficiency centrifugal chillers, high-efficiency fans equipped with efficient motors and variable speed drives. Operational choices are also important: optimal thermostat settings and ventilation can significantly reduce cooling energy use.

Energy efficiency improvement studies conducted with and for customers in the modern service sector in South-East Asia have shown that in new commercial buildings 30 to 40 per cent of the electricity consumption can be saved using presently available technologies. The calculated payback for retrofit investments varies between one and four years, and could thus be attractive for the management of commercial buildings.

Restraining the growth of electricity consumption in modern commercial buildings is one of the most important and attractive opportunities to realize the energy efficiency potentials.

Gradual improvement in energy efficiency of household appliances can be expected to occur as households become able to afford better quality devices. Traditional bioenergy generation and end-use methods are very energy-inefficient and the transition to fossil fuels/electricity will continue, contributing to an increase in both useful as well as primary energy consumption. At the same time, most households in the developing countries will demand higher levels of services, especially when these will be met by the expansion of electricity grid. Hence, average energy use per device may increase even if the technical energy efficiency improves.

In cooking, energy can be conserved in rural areas by introducing more fuel-efficient biomass stoves. Energy efficiency will also continue to increase with the continuing transition to kerosene and LPG stoves. In some countries, LPG stoves presently in use may be improved.

Lighting can be made energy-efficient through the use of circular fluorescent lamps and compact fluorescent lamps (CFLs). The country studies on Brazil and India have shown that the large-scale introduction of CFLs can be very cost-effective from national and utility perspectives. [10] However, to this day the high first costs of CFLs on the one side, and subsidized electricity prices on the other side, limit the market penetration of energy-efficient lighting.

In many developing countries, air-conditioning used to be only a minor energy end use. However, with growing prosperity air-conditioning will be more widely used in a growing number of homes, in particular in countries with warm and humid climates. Air-conditioners vary widely in their energy efficiency. Adequate designs of new buildings and improved insulation can be used to reduce cooling needs.

Water heaters are also gradually becoming more common in a growing number of developing countries. Again, energy efficiency varies with products and with the technology used. Whereas storage water heaters are often energy-intensive, hot water heat pumps are more energy-efficient and may be cost-effective.

Residential space heating is not a major end use in developing countries, except for China, where coal is used to heat homes, and the Republic of Korea. There are considerable potentials to improve heating equipment efficiency and building thermal integrity. However, higher energy efficiency in space heating can only be expected to be achieved gradually over a long period of time.

D. Essential elements of policies and programmes for energy efficiency promotion

The existence of large and cost-effective opportunities for energy end-use conservation in buildings, transport, industries and businesses raises the question why such cost savings have not been realized. Throughout the Asian region some barriers are common to saving energy, such as lack of information or awareness among consumers and institutional barriers. The analysis of barriers to energy efficiency is given considerable attention in the energy efficiency literature.[11] Figure V also attempts to present an overview of the many factors preventing the achievement of energy efficiency.

[10] A. Gadgil and G.Jannuzzi, *Conservation potential of compact fluorescent lamps in India and Brazil,* Energy Policy, (1991), pp.449-463.

[11] A.K.N. Reddy, *Barriers on Improvements in Energy Efficiency,* Energy Policy, 19(10), (1991), pp.953-961; Sioshansi, op. cit.; Schipper and Meyers and others, op. cit.

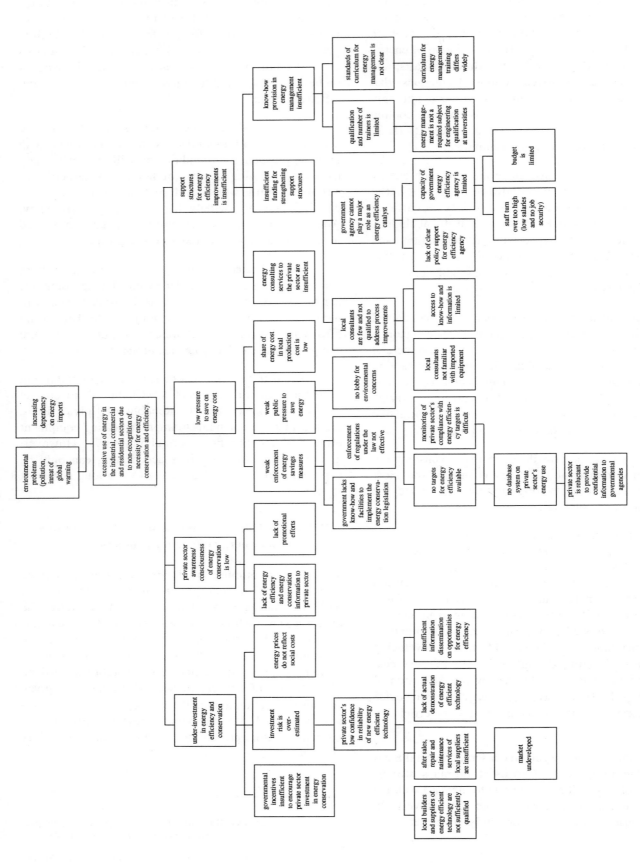

Note: This figure was adapted from a similar figure developed by participants in the "Energy efficiency promotion project" planning workshop organized by the Department of Energy Affairs (now known as Department for Energy Development and Promotion), with the assistance of the German Technical Cooperation Agency (GTZ) in Bangkok, 1992

Figure V. Barriers to achievement of energy efficiency: problems and their roots

A large number of investors and a majority of the consumers have insufficient information on available and rapidly evolving technologies and are uncertain regarding savings and cost-effectiveness. Related barriers include lack of capital and resistance to buying equipment at greater purchase cost (certainly in low-income households), and requirements for rapid payback for building owners, occupants, industries and businesses. In comparison with the industrialized world, capital costs in developing countries are comparatively high, as opposed to energy and labour costs. For the developing countries of the Asian region, in particular among the small and medium-scale industries, the initial prevailing preference is for less capital-expensive (but often not energy-efficient) technologies, equipment and products. Sometimes, responsibilities for making capital investment (e.g. the air conditioner in a building) and paying operating cost (e.g. paying the electricity bill of that air conditioner) are separated. Often, the construction sector tends to compromise on the energy efficiency of buildings to minimize construction costs and maximize sales and profits. Therefore, consumers tend to prefer the cheaper (and often less energy-efficient) technologies, equipment and products, without adequately taking into account, or without being able to take into account, life cycle operating costs and long-term economic implications, let alone all environmental considerations.

Also, important barriers exist on the institutional and policy side. The international oil prices, which have – in real terms – remained comparatively low since 1986, have contributed considerably to the widespread lack of resource and energy consciousness.

Low-income groups and users of public transport also tend to be very sensitive to oil and energy price increases. The international oil trading community also openly opposes energy taxes for the protection of the environment. Energy policy makers are therefore facing many trade-offs. The prospects for rapid energy price rationalization therefore also depends on the other political objectives which the decision makers will have to take into account.

Thus, government budget allocations for the promotion of energy conservation and efficiency have tended to stagnate or decline in many countries. At the same time, many of the para-statal organizations are prevented from engaging in (semi-) commercial energy efficiency promotion. Power plant suppliers and owners are often powerful vested interests, trying to match rising energy demand by supply side measures (capacity expansion) rather than by demand side management.

In many countries of the ESCAP region energy consumption is implicitly subsidized. Often, fuel and electricity prices do not transfer the full costs of energy use to the individual consumers. In particular, electricity tends to be underpriced. In many areas, prices are set by regulations instead of market forces. In particular, electricity pricing is often based on historical accounting costs of utility plants. This practice tends to undervalue the true costs of electricity. Subsidization makes energy consumption cheaper than energy conservation. In addition, time-invariant prices result in too much consumption during peak periods, and too little during off-peak periods.

Hence, all efforts to achieve higher levels of energy efficiency will be fruitless, unless energy pricing systems are adjusted to reflect all costs, including social/environmental costs. Whilst the energy efficiency related literature agrees on this fundamental perspective, it tends to overlook the politicians' dilemma: in many countries, influential pressure groups call for low(er) energy prices. Large-scale industrial and commercial consumers insist on preferential tariff rates.

High import tariffs and non-fiscal barriers may stop energy-efficient products from entering a domestic market. When certain energy-efficient products are manufactured locally, a small market volume may result in high prices. A legislative barrier is lack of efficiency standards, as well as lack of labelling on products.

The medium- and longer-term implications are that the respective economies face a critical and growing risk of structuring their investments and their entire technological systems in ways which may make them become (re) producers of long-term low energy-efficient/high energy consumption techniques. Hence, an effective promotion of energy efficiency appears to be crucial and particularly relevant for the countries of the ESCAP region.

One essential step in improving energy efficiency is the improved flow of information. Available information should be made more accessible and gaps in information should be filled through research and development. Information needed includes improved energy end-use data breakdown, opportunities and magnitude of energy and cost savings of energy-efficient products and processes, local availability and suppliers of such technologies. Information on financing sources and incentives and energy consultants should be included as well. Consumers should be informed about costs and savings of alternatives, and manufacturers should be informed on business opportunities in improving product and process efficiencies.

Dissemination of such information and awareness creation can be promoted by campaigns launched by government agencies and non-governmental organizations, using various media, such as radio, television, brochures, newspapers, etc. Institutional reform may offer some solution: national conservation centres (if transformed into non-governmental organizations) can engage in a greater number of activities of income-generating activities, including training, financing, and informational support. Governments, at the same time, can subcontract such organizations. On the regional level, a systematic and frequent interchange of information and experience among institutions involved in energy conservation will be essential.

The establishment of energy efficiency standards, voluntary or compulsory labelling, building standards and norms and support of activities of consumer protection organizations will go a long way in making more sovereign choices by consumers in their appliance purchases.

To adopt energy conservation, users need an economic incentive. Energy subsidies are often promoted as an instrument for economic growth and to curb inflation, but such relations are highly doubtful, and certainly do not encourage energy efficiency. In the case of electricity, the use of time and seasonal pricing provides incentives for more efficient use. Taxes may provide disincentives to the incorporation of energy-efficient products. Usually taxes are based on its sale price, but duties based on input power would be better from the viewpoint of energy efficiency. Also, important duties could be made dependent on the energy efficiency quality of products.

For many energy efficiency projects, the availability of financing can improve the adaptation of energy conservation. Financing can be channelled through various ways, such as loans through development banks and through electric utilities, and by promoting foreign investment in energy-efficient products and processes. In general, the promotion of international and intra-regional trade in energy efficient products and services will be important to facilitate technology transfer and investment by the private sector in energy efficiency. Training and the support for the formation of commercial energy consulting services and energy service companies may encourage and aid the private sector to develop business opportunities in energy conservation.

In conclusion, essential steps in promotion of energy conservation and efficiency are (a) improving the flow of information to users and manufacturers, (b) consultancy and training (e.g. on energy audits, promotion strategies and energy management), energy standards, labelling and certification, (c) energy servicing, (d) mobilizing domestic and foreign investment in energy conservation, as well as (e) reviewing energy policies (pricing, incentives) tax, and import policies.

International Energy Agency assessment of energy efficiency potential in Organisation for Economic Cooperation and Development countries: summary of opportunities and barriers

	(A) Estimated share of total final consumption	(B) Estimated share of total CO_2 emission	(C) Total energy savings possible	(D) Existing market/inst. barriers	(E) Potential energy savings not likely to be achieved
Residential	11.4%	11.0%	10-50%	Some/Many	Mixed
Space heating and conditioning residential	3.4%	3.6%	Mixed	Some/Many	Mixed
Water heating residential refrigeration	1.1%	2.1%	30-50%	Many	10-30%
Residential lighting	0.6%	1.2%	Over 50%	Many	30-50%
Commercial space heating and conditioning	6.1%	6.8%	Mixed	Some/Many	Mixed
Commercial lighting	1.5%	3.4%	10-30%	Some/Many	Mixed
Industrial motors	4.5%	9.0%	10-30%	Few/Some	0-10%
Steel	4.1%	4.6%	15-25%	Few/Some	0-15%
Chemicals	8.4%	5.9%	10-25%	Few/Some	0-20%
Pulp and paper	2.9%	1.2%	10-30%	Few/Some	0-10%
Cement	0.1%	0.9%	10-40%	Few/Some	0-10%
Passengers cars	15.2%	13.7%	30-50%	Many	20-30%
Goods vehicles	10.1%	9.1%	20-40%	Some	10-20%

Source: International Energy Agency (IEA), *Energy Efficiency and the Environment,* (Paris, 1991), p.26 and p.150.

How to read this table: For example, for lighting, over 50 per cent per unit savings would result if the best available technology were used to replace the average lighting stock in use today over the next 10 to 20 years. Some of these savings would take place under existing market and policy conditions. But due to the many market and institutional barriers, there would remain a 30-50 per cent potential for savings that would not be achieved.

1.   Based on a comparison of the average efficiency of existing capital stocks to the efficiency of the best available new technology. This estimate includes the savings likely to be achieved in response to current market forces and government policies as well as those potential savings (indicated in column E) not likely to be achieved by current efforts.

2.   Extent of existing market and institutional barriers to efficiency investments.

3.   Potential savings (reductions per unit) not likely to be achieved in response to current market forces and government policies (part of total indicated in column C).

4.   Energy use only.

CHAPTER VIII

ENVIRONMENTAL IMPACT OF PRODUCTION AND USE OF ELECTRICITY IN THE ASIAN AND PACIFIC REGION*

Introduction

Electricity demand and production in the Asian and Pacific region has been increasing at a much higher rate than total energy consumption. Whereas in the 1980s overall energy consumption increased at an average annual rate of 4.45 per cent, electricity production increased at 6.6 per cent during the same period. In developing economies the situation presented even more of a contrast, with significantly higher growth in electricity production (over 9 per cent) as opposed to less than 6 per cent growth in total commercial energy consumption. These high growth rates of both energy and electricity consumption are due to the region's rapid economic growth coupled with its high population growth. With much demand as yet unmet, and consequent per capita energy and electricity consumption being at a very low level, as compared with industrialized economies, the high growth in electricity demand is expected to continue. Electricity, as a high-grade and clean form of energy, has many advantages in comparison with other forms of energy. Therefore, electricity production and consumption will continue to increase at a faster rate than overall energy demand.

The share of primary energy use in electric power generation has also been increasing steadily over the years. As of 1990, the share of primary energy used for electricity generation out of the total energy supply had reached a range of between 27 and 41 per cent in various subregions of Asia and the Pacific.

As one of the largest energy-consuming sectors, the electric power industry has environmental and health impacts, particularly in relation to the production of electricity, which has emerged as one of the most significant issues in the development and management of the power system. Many environmental pollutants can be controlled within acceptable limits but unfortunately at a high cost which in some cases is prohibitive.

Energy sources used for electricity generation include coal, oil, natural gas, hydropower, nuclear and to a lesser degree other forms of new and renewable energy, such as geothermal, wind and solar energy. In addition to the environmental impacts of mining, transport, and handling or harvesting these primary energy sources, environmental pollution takes place in their conversion to as well as use of electricity.

Types of pollution depend on the fuel used and types of powerplants. In general almost all plants cause some degree of land, water and air pollution. Other pollution includes loss of species' diversity, population displacement, effects on marine life, radiation and noise.

Although electricity use does not cause any direct environmental pollution, its wasteful use necessitates more production, thereby causing added environmental pollution. Therefore, efficient use of electricity and system loss reductions are considered effective means of reducing pollution. In order to reduce CO_2 emission, these are the only effective means (in addition to fuel switching).

I. OVERVIEW OF THE POWER SECTOR

In another note by the ESCAP secretariat, "Prospects for efficient utilization and load management in electric power utilities" (NR/PCESD/7), reproduced in the present publication, a review was made of installed generating capacity, generation, consumption, efficiency, system losses and future perspective of the power sector in Asia and the Pacific.

II. STRUCTURE OF THE ELECTRIC POWER INDUSTRY

In the ESCAP region as a whole, the power industry is dominated by thermal power capacity as well as thermal generation. As of 1990, the share of thermal power capacity was 71 per cent and that of thermal generation 72 per cent. The breakdown of the installed generating capacity and the generation in 1990 based on types of prime mover or energy sources are given in tables 1 and 2. The situation in the developed and developing economies is similar except that the shares of thermal power were higher in developing countries by a few percentage points. The figure below is the graphical

---

* Note by the ESCAP secretariat for the Expert Group Meeting Preparatory to the First Session of the Committee on Environment and Sustainable Development, Bangkok, 30 September - 2 October 1993 (NR/PCESD/6)

Table 1. Generating capacity breakdown, 1990
(MW)

	ESCAP region	Percentage	ESCAP developed countries	Percentage	ESCAP developing countries	Percentage
Primary	161 934.00	29.30	78 749.00	33.80	83 185.00	26.10
Hydro	122 223.00	22.10	49 616.00	21.30	72 607.00	22.80
Nuclear	38 043.00	6.90	28 862.00	12.40	9 181.00	2.90
NRSE[a]	1 668.00	0.30	271.00	0.10	1 397.00	0.40
Secondary (Thermal)	390 009.00	70.70	154 455.00	66.20	235 554.00	73.90
Steam	347 032.00	62.90	146 622.00	62.90	200 410.00	62.90
Internal combustion	22 185.00	4.00	3 189.00	1.40	18 996.00	6.00
Gas turbine	20 792.00	3.80	4 644.00	2.00	16 148.00	5.10
Total	551 943.00	100.00	233 204.00	100.00	318 739.00	100.00

Source: ESCAP, *Electric Power in Asia and the Pacific 1989 and 1990* (ST/ESCAP/1286).

[a] NRSE: Generating capacity based on new and renewable sources of energy including geothermal but excluding medium and large hydro.

Table 2. Generating breakdown, 1990
(GWh)

	ESCAP region	Percentage	ESCAP developed countries	Percentage	ESCAP developing countries	Percentage
Primary	651 609.00	27.80	335 729.41	32.70	315 879.84	24.00
Hydro	379 429.64	16.20	133 456.41	13.00	245 973.23	18.70
Nuclear	261 301.00	11.20	202 272.00	19.70	59 029.00	4.50
NRSE[a]	10 878.00	0.50	1.00	0.00	10 877.61	0.80
Secondary (Thermal)	1 689 115.79	72.20	690 988.30	67.30	998 127.06	76.00
Steam	1 599 641.52	68.30	670 763.42	65.30	928 878.20	70.70
Internal combustion	44 575.96	1.90	9 556.24	0.90	35 019.72	2.70
Gas turbine	44 898.31	1.90	10 669.07	1.00	34 229.24	2.60
Total	2 340 725.00	100.00	1 026 718.14	100.00	1 314 006.90	100.00

Source: ESCAP, *Electric Power in Asia and the Pacific 1989 and 1990* (ST/ESCAP/1286).

[a] NRSE: Generation utilizing new and renewable sources of energy including geothermal but excluding medium and large hydro.

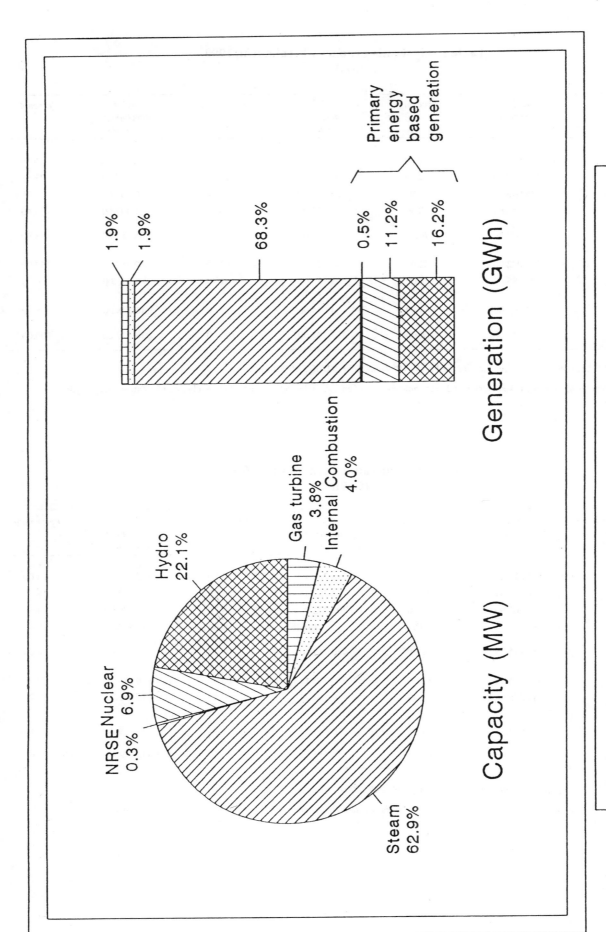

Figure. Share of thermal and primary energy in electric power industry of the ESCAP region, 1990

representation of various shares in the regional context in respect of installed generating capacity and gross generation.

This structure of the power industry is unlikely to change very quickly in the near future. Although inter-fuel substitution within the thermal generation process itself is likely, fuel switching in the past was influenced by the availability of fuel in a country, its national fuel diversification policy and fuel prices. Table 3 shows the trends of primary energy use patterns in electricity generation in selected economies of the ESCAP region. It would be interesting to note how much this trend will change because of environmental concerns in the next few years. There is a movement to switch to environmentally benign fuels such as natural gas; however, this depends very much on the availability of natural gas. Those economies with abundant reserves of natural gas within their national boundaries will probably increase their use at an accelerated pace. A second category of countries (such as Thailand) have already built natural gas-based infrastructures but, with dwindling resources, will probably try to import natural gas from neighbouring countries. A third category of countries, where environmental regulations are more stringent (such as Japan and the Republic of Korea), will probably continue to diversify their generation fuel towards cleaner fuels such as nuclear fuel and imported natural gas in the form of LNG (liquefied natural gas). However, the other major users such as China and India will probably continue to use domestic coal as the major fuel for power generation. Many other developing economies of the region are also opting for more and more use of coal. Table 3 highlights that trend in countries such as Indonesia, Malaysia and the Philippines. The increasing use of coal is almost inevitable because of its availability (reserves), lower price and the rapid progress towards a clean coal technology.

The following chapter examines the environmental implications of the power supply and electricity uses.

III. ENVIRONMENTAL CONCERNS

Increasing concern over environmental degradation affects the electric power industry the most. Although electricity production and use are not the only sources of pollution and greenhouse gas emissions, the power production processes, coupled with the large amount of energy input into them, are significant contributors to land, water and air pollution. Fossil fuel combustion is believed to be responsible for about 60[1] per cent of total CO_2 emissions; of these electricity production accounts for 25-35[2] per cent.

Most of the pollution produced in electricity production and supply comes from the generation process in the thermal powerplants. Hydropower and nuclear powerplants cause different kinds of pollution. Table 4 lists some of the major environmental impacts of electricity supply.

A. Environmental impact of thermal powerplants

In regard to table 2, it may be noted that fossil fuel-based thermal power generation accounts for over 72 per cent of total generation. Therefore this particular area needs greater attention in connection with the formulating of environmental policies and strategies.

The electric power industry is one of the largest consumers (users) of primary commercial energy. Table 5 shows the electricity industry's share in primary commercial energy supply in selected subregions and economies in the Pacific. It is remarkable to note that the share has been increasing since the 1970s. Some highlights are: in the NIEs (newly industrializing economies), the share of the power industry in total primary energy use increased from 24 per cent in 1973 to 34.7 per cent in 1990; in South-East Asia the share rose from 15.6 to 27.4 per cent; in South Asia the share rose from 29.9 to 39.8 per cent; and in the South Pacific the share rose from 20.5 (1980) to 30.8 per cent.

Impacts from powerplant emissions can be classified broadly into three categories: (a) local impacts; (b) regional impacts; and (c) global impacts. There may be some overlaps in the case of certain types of emissions in terms of these groupings.

1. Local impacts

Examples of predominantly local impacts include emissions into the atmosphere from cooling towers, stacks, venting ducts, fuel handling and waste ash disposal activities. They also include effluent discharge into watercourses and waste disposal on land. Hydropower plants cause inundations and population displacements.

---

[1] Advanced Technology Assessment System, issue 7, spring 1992, *Environmentally Sound Technology for Sustainable Development* (United Nations publication, Sales No. E. 92.II.A.6).

[2] *Senior Expert Symposium on Electricity and the Environment: Key Issues Papers* (STI/PUB/899) (Vienna, International Atomic Energy Agency, 1991).

Table 3. Trends of primary energy use in electricity generation
(Tons of oil equivalent and percentages)

Country	Year	Forms of energy use						
		Coal	Crude and petroleum products	Natural gas and LNG	Hydro	Nuclear	Geothermal	Total
China	1980	63 242	20 668	935	15 862			100 707
		(62.8)	(20.5)	(0.9)	(15.8)			(100)
	1985	82 204	14 889	995	25 171			123 259
		(66.7)	(12.1)	(0.8)	(20.4)			(100)
	1989	137 135	20 274	1 145	32 264			190 818
		(71.9)	(10.6)	(0.6)	(16.9)			(100)
India	1980	18 841	1 961	443	16 130	1 040		38 415
		(49.0)	(5.1)	(1.2)	(42.0)	(2.7)		(100)
	1985	37 251	2 697	1 169	18 335	1 792		61 244
		(60.8)	(4.4)	(1.9)	(29.9)	(2.9)		(100)
	1990	47 537	2 680	5 949	21 359	2 179		79 704
		(59.6)	(59.6)	(7.5)	(26.8)	(2.7)		(100)
Indonesia	1980	0	1 779	0	622		0	2 401
			(74.1)		(25.9)			(100)
	1985	576	3 212	0	1 880		48	5 716
		(10.1)	(56.2)		(32.9)		(0.8)	(100)
	1990	3 119	4 054	277	1 269		252	8 971
		(34.8)	(45.2)	(3.1)	(14.1)		(2.8)	(100)
Republic of Korea	1980	682	6 980	0	496	869		9 027
		(7.6)	(77.3)		(5.5)	(9.6)		(100)
	1985	4 150	4 470	0	915	4 186		13 721
		(30.2)	(32.6)		(6.7)	(30.5)		(100)
	1990	4 621	4 658	1 867	1 590	13 222		25 949
		(17.8)	(18.0)	(7.2)	(6.1)	(51.0)		(100)
Malaysia	1980	0	2 346	33	383			2 762
			(84.6)	(1.2)	(13.9)			(100)
	1985	0	2 519	539	1 019			4 077
			(61.8)	(13.2)	(25.0)			(100)
	1990	813	2 989	1 361	915			6 078
		(13.4)	(49.2)	(22.4)	(15.1)			(100)
Pakistan	1980	9	50	1 864	2 074	1		3 998
		(0.2)	(1.3)	(46.6)	(51.9)	(0.0)		(100)
	1985	14	929	2 051	2 913	82		5 989
		(0.2)	(15.5)	(34.2)	(48.6)	(1.4)		(100)
	1990	10	1 597	3 957	4 874	84		10 522
		(0.1)	(15.2)	(37.6)	(46.3)	(0.8)		(100)
Philippines	1980	49	2 692		877		517	4 135
		(1.2)	(65.1)		(21.2)		(12.5)	(100)
	1985	843	1 590		1 383		1 227	5 043
		(16.7)	(31.5)		(27.4)		(24.3)	(100)
	1990	435	2 312		1 509		1 361	5 617
		(7.7)	(41.2)		(26.9)		(24.2)	(100)
Thailand	1980	331	2 942	0	282			3 555
		(9.3)	(82.8)		(7.9)			(100)
	1985	1 142	856	2 388	818			5 204
		(21.9)	(16.4)	(45.9)	(15.7)			(100)
	1990	1 682	2 526	4 183	1 103			9 494
		(17.7)	(26.6)	(44.1)	(11.6)			(100)
Viet Nam	1985	1 560	479	24	428			2 491
		(62.6)	(19.2)	(1.0)	(17.2)			(100)
	1990	1 936	537	21	1 251			3 745
		51.7	(14.3)	(0.6)	(33.4)			(100)

Source: *Energy Indicators of Developing Member Countries of Asian Development Bank* (Manila, Asian Development Bank, July 1992).

Note: Figures in parentheses show percentage share of total energy use.

Table 4. Illustrative environmental impacts of electricity supply

System component	Key impacts
Coal	. Groundwater contamination . Land disturbance, changes in land use and long-term ecosystem destruction . Emissions of SO_2, NO_X, particulates with air quality implications . Heavy metals leachable from ash and stage wastes . Global climatic change from CO_2 emissions . Lake acidification and loss of communities due to acid depositions
Oil and gas	. Marine and coastal pollution (from spills) . Damage to structures, soil changes, forest degradation, lake acidification from S and N emissions . Groundwater contamination . Greenhouse gas emissions impact, e.g. global climate change
Hydroelectric	. Land destruction, change in land use, modification of sedimentation . Ecosystem destruction and loss of species diversity . Changes in water quality and marine life . Population displacement
Nuclear	. Surface and groundwater pollution (mining) . Changes in land use and ecosystem destruction . Potential land and marine contamination with radionuclides (accident conditions)
Renewables	. Atmospheric and water contamination . Changes in land use and ecosystem . Noise from wind turbine operations

Source: Senior Expert Symposium on Electricity and the Environment: Key Issues Papers, (STI/PUB/899) (Vienna, International Atomic Energy Agency, 1991).

Table 5. Electricity's share in primary commercial energy supply (consumption)*

Country or area	1973	1980	1985	1986	1987	1988	1989	1990
Newly industrializing economies	24.0	27.8	32.4	33.4	34.8	34.3	36.0	34.7
South-East Asia	15.6	22.1	28.5	27.1	27.9	28.7	26.5	27.4
South Asia	29.9	35.1	40.7	41.0	39.5	45.2	38.5	39.8
South Pacific		20.5	30.7	32.9	36.1	42.0	35.3	30.8
China		29.6	39.4	29.4	29.6	31.9	35.7	38.6
India	30.7	36.3	42.7	42.8	40.6	47.2	39.2	40.8
Asia Pacific (excl. China)	9.2	29.7	31.8	31.9	32.0	34.4	35.4	37.2
Asia Pacific (excl. China and India)	20.5**	24.7	29.8	30.6	31.8	32.2	32.0	31.9

Source: Energy Indicators of Developing Member Countries of Asian Development Bank, (Manila, Asian Development Bank, July 1992).
*Note: * Electricity share in primary energy refers to ratio of energy used for power generation over primary energy supply.*
* ** 1974 figure.*

2. Regional impacts

Acid depositions from emissions such as SO_2 with residence time of a few days fall into this category which may affect both local areas as well as areas within up to a few thousand kilometres away. Fossil fuel combustion has resulted in significant emissions of SO_2 and NO_x to the atmosphere over recent decades. The sulphur content of coal and oil may be as high as up to 6 and 3 per cent respectively, which on combustion is converted to SO_2 that is emitted through stacks into the atmosphere. The transport of SO_2 over long distances gives rise to the deposition of emission products across national boundaries. The direct effects of SO_2 may result in the degradation of plants' growth, soil acidification and acidification of streams, rivers and lakes. Acidification of water may endanger fish populations whereas both soil and water acidification may affect plantations and forests.

B. Global environmental impacts

A current cause of concern worldwide is the global warming due to the "greenhouse effect" caused by the increase in atmospheric concentration of CO_2 and other radiately active trace gases over the past few decades. Table 6 lists various greenhouse gases with their concentration level (1987) and their annual rate of increase. CO_2 emissions are estimated to account for 50 per cent of the overall greenhouse effect. Because of its potential for global climate change, the issue of greenhouse gas has drawn international attention. Current concerns relate basically to the worry that continued build-up of the greenhouse gas may cause irreversible change in the global climate pattern. In this

Table 6. Greenhouse pages

	Concentration (ppbv)[a]	Annual rate of increase (Percentage)
Carbon dioxide	344 000	0.4
Methane	1 650	1.0
Nitrous oxide	304	0.25
Methyl chloroform	0.13	7.0
Ozone	Variable	–
CFC 11	0.23	5.0
CFC 12	0.4	5.0
Carbon tetrachloride	0.125	1.0
Carbon monoxide	Variable	

Source: ESCAP, *Energy Policy Implicaitons of the Climatic Effects of Fossil Fuel Use in the Asia-Pacific Region* (ST/ESCAP/1007).
[a] Part per billion in volume.

context the main concern is the possible effect of changes in temperature differential between the poles and the equator. This could result in a significant change in wind, rainfall patterns and incidence of storms. Higher temperatures at the poles may also result in melting of the ice caps leading to a rise in sea level.

C. Environmental impacts of hydropower projects

Environmental impacts of a hydropower project involve physical and biological resources, changes in water flow, water quality and marine life, and quality of life including population displacements. The most sensitive issues in large storage-type hydropower plants relate to the alteration of the ecology from that of riverain to a lake regime through building of a dam, inundation of the reservoir area causing destruction of forests and vegetation, displacement of human settlements from the upstream area, endangerment of wildlife and aquatic species, climatic changes, the risk of dam failure, and enhancement or creation of ecological environments favourable to the spreading of parasitic and water-borne diseases.

It may be noted here that some beneficial aspects of water resources development are often overlooked but are important in assessing the overall impact. These benefits include irrigation, fisheries, recreation, navigation and flood control.

D. Environmental impacts of nuclear power plants

The environmental impacts associated with nuclear power production are mainly in the form of thermal pollution from the condenser cooling water. This is also the case with any steam powerplants. However, the main concern about the nuclear powerplant is about the possible accidental emissions of radioactive effluents. Another concern is the disposal of radioactive wastes or spent fuel. Although a small amount of radioactive material leakage may be expected routinely from a plant, its collection, treatment and safe disposal ensure that there is no threat to the environment.

E. Asian and Pacific situation and current concerns

A recently held ESCAP workshop concluded in its report that "although in most developing countries the contribution of CO_2 to greenhouse gas emission was not yet a major issue, efforts to improve efficiency so as to contain CO_2 were considered feasible and desirable".[3]

[3] "Report of the Regional Workshop on Environmentally Sound Coad Technologies, Bankok, 7-10 April 1993" (NR/IHE/ESCT/Rep.), para. 49.

This appears to be based on the current low per capita emission from developing economies (table 7). Although the two largest fossil fuel-consuming developing economies in the region, China and India, emitted 610 and 164 million tons of carbon in 1988, their per capita emissions were 0.56 and 0.20 tons respectively. The regional average, which included three developed economies, was 1.2 tons per capita. This compares with 5.34 tons per capita in the United States of America.[4] However, the situation is likely to shift towards more emission from the region owing to higher use of fossil fuel. Table 8 shows some projections for developing economies in the ESCAP region.

Table 7. CO_2 emission from fossil fuel burning and cement manufacture in Asian and Pacific countries, 1988

Country	Total (106/tons)	Per capita (tons/capita)
China	609.9	0.56
Japan	269.8	2.2
India	163.8	0.20
Australia	65.8	4.0
Republic of Korea	55.8	1.3
Indonesia	37.8	0.22
Thailand	17.9	0.33
Pakistan	15.9	0.14
Malaysia	11.9	0.72
Philippines	10.8	0.18
Viet Nam	5.1	0.08
Bangladesh	3.7	0.03
Brunei Darussalam	1.3	5.2
Sri lanka	0.95	0.06
Papua New Guinea	0.60	0.16
Nepal	0.27	0.01
Cook Islands	0.16	7.9
Fiji	0.15	0.20
Lao People's Democratic Republic	0.06	0.01
Maldives	0.02	0.12
Regional total	5 893.00	1.2

Source: Gregg Marland, "The impact of energy systems on atmospheric carbon dioxide", in *Burning Questions,* K.V. Ramani and others, eds. (Kuala Lumpur, Asian and Pacific Development Centre, 1992).

---

[4] Gregg Marland, "The impact of energy systems on atmospheric carbon dioxide", in *Burning Questions,* K.V. Ramani and others, eds. (Kuala Lumpur, Asian and Pacific Development Centre, 1992).

Another conclusion of the same workshop is also very relevant for this analysis. It says that "other pollution problems arising from SO_x, NO_x and waste disposal were considered severe in many instances and required immediate attention". These conclusions apply to thermal power generation.

Considering the fact that the current situation in various countries in the region would be different in respect of needs and options to address environmental issues, it is not possible to find a regional common strategy. Generally many of the pollutants can be controlled by different optimal combinations of technological applications in the pre-combustion, combustion and post-combustion phases of the fossil fuel chain. For example, the sulphur and ash content of coal can be reduced through beneficiation (washing). This is particularly effective when coal has to be transported over a long distance. Combustion technology has also advanced remarkably during the last few years. Given the fact that many existing fossil-fired power stations will continue to be used for some years, it may be desirable to look at the techno-economic potential of refurbishment and retrofitting.

Although pollution control technologies such as flue gas desulphurization and $deNO_x$ equipment are available, they add significantly to plant costs in the order of 25 to 30 per cent of the project cost. Fluidized bed combustion (FBC) and an integrated gasification combined cycle (IGCC) can minimize some of the pollutant emissions.

In the area of hydro and nuclear power generation, the current situation in Asia and the Pacific is dominated by public perceptions of the problem rather than by facts in many cases. What is missing is an objective assessment of the real situation. It is dovious that any project will have some negative impact on the environment. Therefore a kind of trade-off has to be made based on the findings of the evaluation on a case-by-case basis. Many of the concerns can be addressed with satisfactory results if properly planned and if adequate measures are taken from the very beginning of a project.

IV. INTEGRATION OF ENVIRONMENTAL CONCERNS INTO POWER SYSTEM POLICY AND PLANNING

It is well understood that electricity production and use entail some adverse impacts on the environment. The process of the integration of environmental concerns into power system policy and planning involves the

Table 8. Actual and projected total carbon emissions in selected developing economies of the ESCAP region

(All figures in millions of tons of carbon)

Country/area	1986	2000		2010		
	Actual	S1	S2	S1	S2	S3
GROUP A						
China	548.90	937.50	789.70	1 353.40	1 047.20	993.60
Democratic People's Republic of Korea	35.50	53.70	43.00	72.40	50.70	44.00
India	119.12	225.62	199.80	394.75	298.89	277.17
Indonesia	26.98	67.08	60.40	127.30	99.23	95.52
Islamic Republic of Iran	36.02	61.91	54.86	86.83	67.10	65.72
Republic of Korea	47.90	99.10	89.70	126.90	106.50	93.40
Taiwan Province of China	23.50	49.20	39.90	63.20	48.30	44.80
GROUP B						
Bangladesh	3.30	5.80	5.00	8.10	6.30	6.10
Hong Kong	7.50	11.00	9.60	16.40	13.40	9.90
Malaysia	8.85	15.82	14.19	29.27	22.73	21.32
Pakistan	13.10	31.00	26.50	48.90	37.70	35.20
Philippines	6.96	14.42	12.77	24.16	18.05	17.38
Singapore	10.04	14.55	13.49	19.04	16.02	15.67
Thailand	12.16	41.75	36.84	93.74	71.94	65.75
Viet Nam	4.48	8.88	8.43	14.49	13.04	13.04
GROUP C						
Afghanistan	0.07	1.60	1.40	2.70	2.10	2.00
All other South-East Asian countries	3.06	6.31	6.00	10.23	9.20	9.20
Maldives	0.00	0.10	0.10	0.10	0.10	0.10
Nepal	0.20	0.40	0.40	0.60	0.50	0.40
Pacific islands	2.10	3.50	3.00	4.90	3.80	3.40
Sri Lanka	0.90	2.80	2.40	4.30	3.30	2.50
TOTAL (A+B+C)	911.25	1 652.04	1 417.49	2 501.71	1 936.12	1 816.18

Source: ESCAP, *Energy Policy Implications of the Climatic Effects of Fossil Fuel Use in the Asia-Pacific Region* (ST/ESCAP/1007).

Notes: S1 = Base-case scenario (all countries).

S2 = Energy-efficiency scenario (all countries).

S3 = Low-carbon scenario (all countries).

assessment and valuation of these impacts and then control and minimization of the adverse effects. The whole chain of the electricity production process including the primary energy supply source as well as the generation, transmission and distribution should be analysed and taken into account in identifying environmental and health impacts and ways and means to mitigate negative impacts. If electricity can be generated by more efficient and cleaner processes, substitution of electricity for other forms of fuels would make a positive contribution towards the protection of the environment. The process of the integration of environmental concerns starts with the planning process. The planning and management of the electric power industry should be carried out through an integrated approach so that the links between electricity demand and supply and the environmental impacts and other socio-economic parameters can be harmonized in an optimal manner.

Traditionally power development decisions were based on the least cost supply options and mainly on internal cost comparisons of the different fuels. The integrated approach incorporates all the costs, including the social, environmental and health effects. This is what is called the internalization of the costs and benefits of the environmental and health impacts of electricity production and supply processes. In considering the non-quantifiable environmental, social and economic impacts, a multi-criterial approach in the decision-making process can be applied in making a choice among several options.

The electricity sector is a subsector of the energy sector, which itself is a component of the overall socio-economic system. Each component and subcomponent has its own objectives which often conflict with those of other components. To satisfy the overall socio-economic objectives a trade-off is essential. This is particularly true in the case of environment and energy development. Intensive and continuous interaction between specific electricity plans in the context of overall energy strategy and overall socio-economic planning is needed to avoid or at least minimize conflicts between decisions or objectives.

Although a number of techniques have been developed for valuing health and environmental costs, this is a difficult task because of the uncertainties involved in some of the valuations. Some of the previously ignored inputs and outputs such as environmental impacts should be quantified and valued. An alternative means of incorporating health and environmental costs is to impose additional constraints on the system optimization process. Special difficulties are involved in applying this process to the regional or global level. Some environmental concerns, such as emissions of CO_2, severe accidental release of radioactive materials from a nuclear plant and acid rain, are transboundary in nature. Their incorporation in the decision-making process requires global and regional approaches including regional and international cooperation.

Economic efficiency is the guiding criterion in electricity planning. Socio-political objectives may also be addressed by adding constraints to the system optimization model. Economic efficiency must take into account the costs of depleting natural resources, including land use, water or air pollution and their consequences. Setting of electricity tariffs to reflect some of these impacts is a major issue, particularly in developing economies, where other objectives such as purchasing power and equity come into play.

An environmental impact assessment (EIA) should be carried out in all phases of power system development to determine what technology is appropriate to mitigate any negative impact. It is advisable to incorporate environmental concerns in policy, planning, development and management of electric power systems to ensure a consistent approach to mitigate negative impacts.

The important first step towards the integration of environmental concerns into energy and electricity planning is to work towards developing acceptable environmental standards on various types of pollution with a time-frame to achieve those standards. The next step would be to incorporate gradually those standards into the electricity sector's decision-making. The decisive factor in setting and meeting the target is cost. The other associated issue is technology. Table 9 shows the standards and targets set by the Republic of Korea.

Table 9. An example of emission standards applicable in the Republic of Korea

Current SO_2 and NO_x until December 1994

Emission	Fuel source	Limits	Current status
SO_2 (PPM)	Anthracite coal	1 200-1 650	740-1 560
	Bituminous coal	700	230-470
	Heavy oil	1 200	110-1 120
NO_x (PPM)	Coal	350	110-330
	Heavy oil	250	130-240

Future standards in the Republic of Korea, 1995-1999

Emission	Fuel source	Limits	
		Jan. 1995-Dec. 1998	After Jan. 1999
SO_2 (PPM)	Anthracite coal	1 200-1 650	270
	Bituminous coal	500	270
	Heavy oil	1 200	270
NO_x (PPM)	Coal	350	350
	Heavy oil	250	250

V. ALTERNATIVE APPROACHES TO ADDRESS ENVIRONMENTAL ISSUES

It is generally agreed that there are mainly three basic ways to reduce and/or control environmental degradation from the energy and power industry. These are:

(a) Efficiency improvement both in the power industry itself in the area of operation including reducing system losses and in end-use (energy intensity);

(b) Switching to cleaner fuel;

(c) Pollution control technology.

These approaches are not mutually exclusive.

The efficiency measure can be applied by using best practices and best technology in the production, supply and consumption of electricity. The aspects of efficiency have been reviewed in greater detail in two other notes by the ESCAP secretariat "Prospects for energy conservation and efficiency in the ESCAP region" and "Prospects for efficient utilization and load management in electric power utilities" reproduced in the present publication.

The second measure of fuel switching from high pollutant fuels (such as coal and oil) to less polluting fuels (such as natural gas and nuclear power) and further to new and renewable energy (such as hydro, solar or wind energy) is a longer-term policy measure. It involves pricing and non-pricing measures. Pricing measures include tariffs reflecting the true cost of electricity produced and perhaps some kind of environmental taxes that would encourage the power industry to use cleaner fuels. Non-pricing measures include regulation and/or acts specifying or limiting emissions.

Pollution control technologies such as particulate emission controls, flue gas desulphurization and other pre-combustion, combustion and post-combustion technologies or clean fuel technology can be considered control measures. Combustion technologies such as FBC, IGCC also fall into this category. In many cases application of these technologies adds significantly to the cost of generation and reduces efficiency (because of its own energy consumption). Given the current and projected power development trends in the region (thermal power domination), the role of technology should be significant. This is also important because many existing coal- or oil-fired boilers will be in use for a long time into the future. There is, therefore, scope for refurbishing or retrofitting of these plants with more efficient pollution abatement technologies.

VI. CONCLUSIONS AND RECOMMENDATIONS

The following conclusions and recommendations are hereby drawn:

1.   The energy/electricity demand in the ESCAP region, particularly in developing economies, will increase at a high rate until the year 2000 and beyond though at a gradually diminishing rate.

2.   It is evident that the power industry will continue to use a high proportion of fossil fuel for electricity production and the environmental impacts of fossil fuel burning in the region will therefore increase. While air pollution, such as SO_2, NO_x, and particulate emissions, is controllable, though at a cost, through clean fuel technology the CO_2 emissions are likely to increase in the region.

3.   The current state of technology in the developing economies' power industry implies low operating efficiency, high system losses and resulting high per unit environmental pollution. Measures to improve these parameters would automatically reduce per unit pollution. Fuel switching is not expected to be significant before the end of the century.

4.   Some pollution control technologies can be adopted to reduce negative environmental impacts.

5.   It is desirable to set environmental standards and take appropriate measures to control pollutions.

6.   On regional and global environmental issues, regional cooperation in taking concerted steps in needed.

7.   The issue of transfer of technology and funding support should be addressed by all parties concerned including the developer of the technology and the potential users. The international community can play a facilitator's role in the dialogue.

CHAPTER IX

PROSPECTS FOR EFFICIENT UTILIZATION AND LOAD MANAGEMENT IN ELECTRIC POWER UTILITIES[*]

Introduction

To meet sustained high economic growth, the demand for electricity in developing economies of the ESCAP region has been growing at a remarkably high rate and this growth is expected to continue in the future.

Over the years the electric power industry has become increasingly complex and capital intensive. Many utilities in developing economies have seen phenomenal growth in the recent past, which sometimes resulted in deficiency in terms of manpower and managerial capability. This situation has often led to a sub-optimal system with many technical and managerial problems. In many developing economies, inappropriate pricing and non-pricing policies on energy and electricity use have frequently resulted in inefficient production and wasteful usage of energy and electricity. Much scope exists for better management of the energy/ electricity sector, which could lead to economic efficiency, energy efficiency, energy conservation and environmental protection.

To improve energy utilization it is essential to optimize the supply and manage the demand. Optimizing the supply involves least cost investment planning to meet the forecasted demand and the optimal operation of the system. Plant performance, including improved capacity utilization, load factor improvement, efficient operation and maintenance and the reduction of system losses, is an essential element of this optimization.

Greater attention ought to be given to the efficient utilization of the power system. In many utilities, operating efficiencies are so low and system losses are so high that the prospects are good for improving those system parameters that could create additional capacities or at least relieve a part of the demand for capacities for a long time. The demand management activities include pricing and customers' load control, as well as end-use efficiency improvements.

The utilities need to keep the balance between supply and demand at all times, through adjustments in supply and/or demand in a consistent and reliable manner. Traditionally, electric power utilities have been managing the supply side to match the demand. The concept of demand side management (DSM) as a total energy savings approach is being increasingly accepted as an approach to reducing peak as well as base load to create a long-term structural energy efficiency improvement in the economy. DSM is a process of identifying and implementing initiatives that improve the use of energy supply capacity by influencing the timing, amount and characteristics of the demand for electricity by customers. A DSM programme is recognized as a resource in the integrated resource planning (IRP) in modern power system development strategy.

In this note, an attempt is made to elaborate on the above issues, after first analysing the current situation in the electric power industry in the region, in order to examine the prospects for efficient utilization of the resources of the utilities and for the possible adoption of appropriate load management methods and technologies.

I. CURRENT STATUS OF THE ELECTRIC POWER INDUSTRY IN THE ASIAN AND PACIFIC REGION

The size and structure of electric power utilities in the Asian and Pacific region vary widely from country to country. Capacity in 1990 ranged from 1.2 MW in Niue to almost 192,000 MW in Japan. The demand and supply situation, trends, problems and development prospects of the power supply industry can therefore vary considerable from one country to another.

Australia, Japan and New Zealand are the developed economies within the ESCAP region; all others are developing economies. These two groups must therefore be analysed separately. In the region as a whole, until the early 1970s, electricity generation showed a steady high average growth of over 9 per cent a year. But after the sudden rise in oil prices in 1973, growth slowed down to 8.7 per cent and further to 6.6 per cent in the 1980s. During the second half of the 1980s, however, growth showed an upward trend, with an average annual rate of 7.3 per cent.

[*] Note by the ESCAP Secretariat for the Expert Group Meeting Preparatory to the First Session of the Committee on Environment and Sustainable Development, Bangkok, 30 September - 2 October 1993 (NR/ PCESD/7).

Figure I shows the electricity generation trends during the period 1970-1990. Because the data were not available, the total for the region and for the developing countries up to the year 1978 does not include China. As expected, the trends in developed and developing countries were not the same. Whereas in developed countries, the consequences of the two oil price rises were felt immediately in electricity generation, in developing countries, the effects were somewhat delayed and less prominent. Except for a slight drop in 1975, there was steady growth in electricity generation in developing countries throughout the whole period. In developed countries growth in generation suffered an initial set-back but was to recover, though to a lower level, some years later. The pattern varied widely from one country to another in both developed and developing economies.

The relative share of electricity generation in the ESCAP developing economies, which since 1985 has exceeded that of the ESCAP developed economies, further increased to over 56 per cent at the end of 1990. This was attributable to low per capita electricity consumption in developing countries, coupled with the low access level to electricity in many of the countries. Their demand for electricity is expected to grow at a much higher rate than that of the developed countries for many years to come. It is likely, therefore, that the above gap will continue to widen, representing an increasingly larger share of developing economies in the total electricity production of the region.

In the region as a whole, the share of generation by thermal plants rose from 56.2 per cent in 1962 to a peak of 77.6 per cent (including gas turbines and diesel power plants: 1.5 per cent) in 1973. There was subsequently a significant downward trend to 69.6 per cent in 1986. However, owing to the oil price crash in 1986 and the continued depressed oil market and resulting other energy price market, the trend has since reversed towards more thermal generation. In 1990, the share of thermal generation was 72.2 per cent.

Figure II shows the status of electricity generation in the region, by class of generator prime movers.

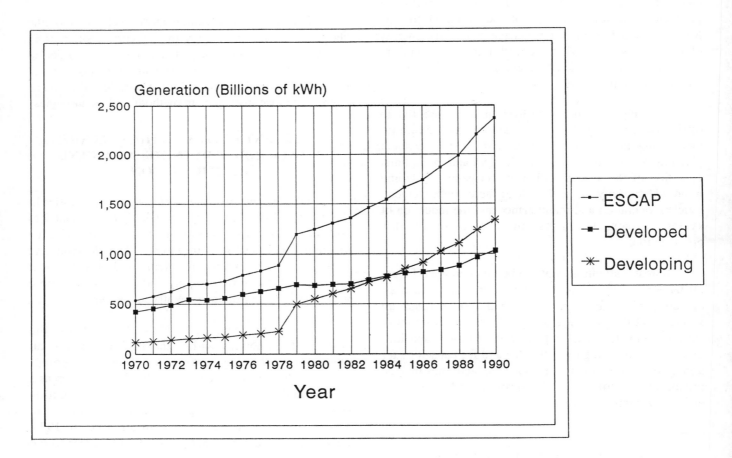

Figure I. Electricity generation, ESCAP region, 1970-1990

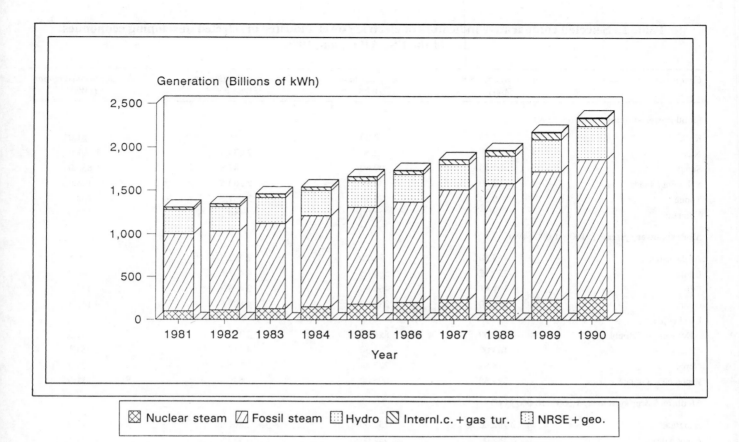

Generation (Billions of kWh)

Legend: ☒ Nuclear steam ▨ Fossil steam ⊞ Hydro ◹ Internl.c. + gas tur. ⊡ NRSE + geo.

Figure II. Generation by class of prime movers in the ESCAP region, 1981-1990

Table 1 shows a comparison of key indicators of electricity generation, consumption and capacity utilization in selected developing economies of the ESCAP region in 1990.

Table 2 shows to a similar comparison of selected developed economies of the world, including the three developed countries of the ESCAP region.

II. GROWTH PERSPECTIVES FOR THE POWER INDUSTRY IN RESPECT OF DEMAND AND ENERGY IN THE REGION

As the economy grows, so will demand for electricity. The per capita electricity consumption level in most of the developing economies is very low, as can be seen from table 1. Except for Guam, Macau, Hong Kong, the Republic of Korea and Singapore, none of the developing economies of the ESCAP region had achieved the 1990 world average per capita consumption of 2,207 kWh (tables 1 and 3). Because most of the economies in the region are developing, the regional average per capita electricity consumption (699 kWh) was less than one third of the world average. The average per capita consumption (398 kWh) of developing economies of the region was a little over half

the regional average and less than one fifth of the world average.

The situation varies so widely among countries that per capita consumption in a rich economy is a few hundred times that of a poor one, mainly because only a small fraction of the population (most of whom live in rural areas) have access to electricity. In an Asian Development Bank report[1] it has been estimated that in a number of Asian developing economies no more than 25 per cent of the population have access to electricity. There is a considerable amount of unmet demand in most of the developing countries, constrained by inadequate supply. The demand, therefore, is almost certain to grow at a fast rate.

A World Bank survey[2] revealed that most developing economies had ambitious plans for developing electric power. During the 1990s, the

---

1 Asian Development Bank, *Energy Indicators of Developing Member Countries of Asian Development Bank* (Manila, Asian Development Bank, 1992).

2 *Capital Expenditures for Electric Power in the Developing Countries in the 1990s,* the World Bank Industry and Energy Department Working Paper, Energy Series Paper No.21 (Washington D.C., The World Bank, 1990).

Table 1. Selected comparative indicators of electric power industry in selected developing economies of the ESCAP region, 1990

Country or area	Installed capacity (MW)	Gross generation (GWh)	Capacity utilization ratio (kWh/kW)	Per capita consumption (kWh)
Small power systems (up to 100 MW)				
Kiribati	3.70	8.20	2 216.2	81.0
Niue	1.20	2.50	2 083.3	1 109.0
Samoa	29.40	99.60	3 387.8	1 835.0
Solomon Islands	17.00	38.48	2 263.5	108.0
Tonga	6.84	20.98	3 067.3	202.3
Vanuatu	9.37	23.90	2 550.7	150.0
Modest power systems (>100 – 1 000 MW)				
Afghanistan	453.50	1 121.20	2 472.3	49.5
Bhutan	352.77	1 564.05	4 433.6	89.8
Fiji	191.41	466.90	2 439.3	497.2
Guam	282.00	1 381.67	4 899.5	9 122.3
Lao People's Democratic Republic	150.00	820.32	5 468.8	41.3
Macau	183.00	753.50	4 117.5	2 264.1
Nepal	278.82	773.84	2 775.4	29.3
Papua New Guinea	395.30	605.00	1 530.5	151.0
Medium power systems (>1,000 – 10 000 MW)				
Bangladesh	2 702.00	7 732.00	2 861.6	43.0
Hong Kong	8 387.00	28 984.00	3 455.8	4 574.5
Malaysia				
Peninsular	5 040.00	21 285.00	4 223.2	1 236.7
Sabah	303.30	1 083.65	3 572.9	n.a.
Sarawak	363.43	1 398.29	3 847.5	500.0
Myanmar	1 091.20	2 477.56	2 270.5	45.3
Pakistan	6 409.00	31 427.00	4 903.6	265.0
Philippines	6 356.80	25 249.00	3 972.0	379.0
Singapore	3 690.80	15 714.00	4 257.6	5 283.0
Sri Lanka	1 289.65	3 150.00	2 442.5	154.0
Thailand	9 722.0	44 175.00	4 543.8	681.0
Large power systems (> 10,000 MW)				
China	137 890.00	621 318.00	4 505.9	458.9
India	74 456.00	288 709.00	3 877.6	253.4
Indonesia	18 299.00	59 244.53	3 237.6	154.7
Iran (Islamic Republic)	17 952.00	59 102.00	3 292.2	853.0
Republic of Korea	24 056.00	118 461.00	4 924.4	2 453.0
Total	320 362.49	1 337 190.17	4 174.0	397.8

Source: United Nations publication, Electric Power in Asia and the Pacific, 1989 and 1990.

Table 2. Selected comparative indicators of electric power industry in selected developed economies of the world, 1990

Country	Installed capacity (MW)	Gross annual generation (GWh)	Capacity utilization ratio (kWh/kW)	Per capita consumption (kWh)
ESCAP region[a]				
Australia	34 434	143 233	4 159.6	7 255
Japan	191 945	857 272	4 466.2	6 197
New Zealand	7 185	30 853	4 294.2	8 116
Other regions[b]				
Canada	104 140	481 791	4 626.0	18 149
France	103 410	419 584	4 057.0	6 661
Federal Republic of Germany (Former)	99 750	454 710	4 558.0	7 420
United Kingdom of Great Britain and Northern Ireland	73 059	318 979	4 366.0	5 761
United States of America	775 396	3 031 058	3 909.0	12 170
Union of Soviet Socialist Republic (Former)	333 100	1 726 000	5 182.0	5 856

[a] United Nations publication, Electric power in Asia and the Pacific, 1989 and 1990.

[b] 1990 Energy Statistics Yearbook (United Nations publication, Sales No. E./F.92/XVII.3).

Table 3. Production and consumption of electricity, ESCAP countries or areas compared to world, 1990

Group of countries or areas	Production (GWh)	Consumption Total (GWh)	Per capita (kWh)
World	11 733 858	11 752 243	2 207
ESCAP region	2 368 549	1 947 792	699
ESCAP developed countries	1 031 359	917 382	6 368
ESCAP developing countries or areas	1 337 190	1 080 410	398

Note: This table is based on data from the following publications: *1990 Energy Statistics Yearbook* (United Nations publication, Sales No. E./F.92.XVII.3), New York 1992; and the forthcoming United Nations publication, Electric Power in Asia and the Pacific, 1989 and 1990.

average growth in installed generating capacity and generation has been planned to be 7.5 and 7.0 per cent, respectively, in developing economies in the ESCAP region. This growth rate was, however, much lower than

the average annual growth rate of 9.2 per cent in generation during the 1980s. It is expected that with the surveyed growth rates the installed generating capacity for the region's developing economies would rise to 493,015 MW in 1999 compared to 235,707 MW in 1989. The generation plus net imports would almost double from 1.5 trillion kWh to 2.9 trillion kWh during the same period. Figures III and IV show these trends, including the expected structure of the industry in respect of fuel uses.

III. EFFICIENCY AND ENVIRONMENT

If energy is utilized efficiently, less will need to be produced and used, and less environmental pollution will be caused. Inefficient utilization of energy/electricity in various other sectors of the economy also causes environmental pollution. This situation is further aggravated in low-income economies, which often resort to using less efficient equipment and appliances because the initial investment costs are lower.

The energy services required by the economic sectors are light, heat/cold, and motive power. As efficient utilization of energy/electricity to produce these services will lower the level of environmental pollution, due consideration should therefore be given to environmental factors when energy policies are being formulated. Such policies should strive towards achieving greater efficiency in the production and use of energy and selecting efficient processes and practices that would promote more rational use of energy.

The subject of end-use energy efficiency is now very topical. A separate article in this book has therefore been devoted to addressing this issue in greater detail. In the following sections we analyse the power industry's own efficiency issues in terms of capacity utilization and system losses.

Capacity utilization

In developing the power sector, emphasis is usually placed on adding more and more generating capacity, with inadequate attention being given to full and efficient utilization of the existing capacity. In many developing countries, the operating ratio of electricity generating capacity is well below the regional average of 4,174 kWh/kW for developing countries. As can be seen in table 1, out of 30 developing economy utilities, 11 (or 35.5 per cent) had an operating ratio of below 3,000 kWh/kW. This compares with an average of around 4,000 kWh/kW for the developed economies (table 2). One can also see from the table that except for a few utilities, the capacity utilization of the smaller utilities is

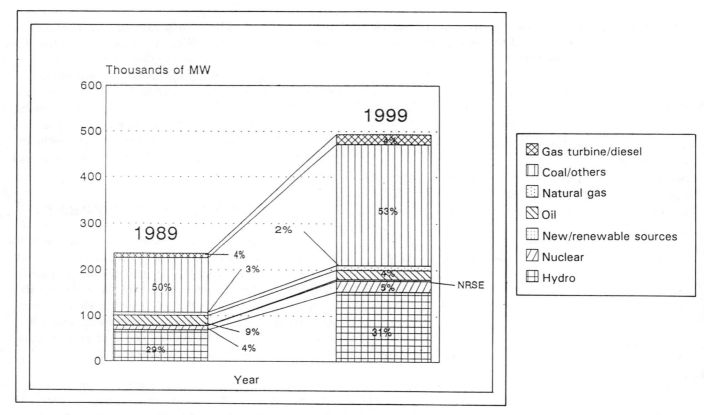

Figure III. Projected installed capacity of ESCAP developing economies (1989 and 1999)

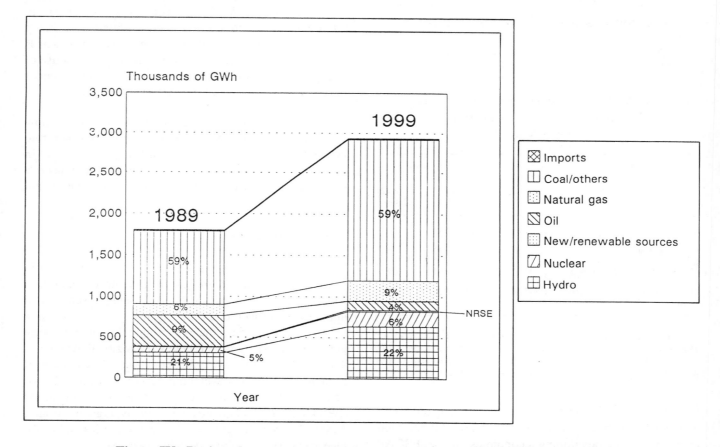

Figure IV. Projected generation of ESCAP developing economies (1989 and 1999)

generally lower. Despite huge unutilized generating capacity, most of those utilities have a relatively high incidence of service interruptions, or what is termed "unreliable power supply". This is mainly due to inadequate transmission and distribution facilities. poor maintenance, lack of spare parts and lack of sustained operating practices.

Power system losses

In supplying electricity to consumers, losses are incurred in the generation, transmission and distribution stages of a power system. Generation losses may vary between 1 and 6 per cent, depending on the type of power station (hydro or coal-fired plant). Recent research suggests that the average energy losses in transmission and distribution should normally be below 10 per cent of gross generation, while economically optimal achievable loss levels may be as low as 5 per cent. Table 4 shows losses in selected economies in the ESCAP region. High transmission and distribution loss is a serious problem for the power utilities of most developing countries in the region. Only a few utilities have succeeded in keeping their losses within reasonable limits.

Table 4. Transmission and distribution losses in selected economies of the ESCAP region, 1990

Economy	Losses (percentage)
Bangladesh	39.1
China (1989)	7.0
Fiji	8.6
Hong Kong	11.4
India	20.7
Indonesia	15.6
Republic of Korea	5.2
Malaysia	13.1
Myanmar	27.3
Nepal	30.8
Pakistan	19.9
Philippines	15.9
Sri Lanka	16.7
Taiwan Province of China	5.9
Thailand	10.6
Viet Nam	27.1
Regional average	11.1

Source: Asian Development Bank, *Energy Indicators of Developing Member Countries of Asian Development Bank* (Manila, Asian Development Bank, 1992).

The results of the analysis of the loss figures are shown in table 5. In 1990 56.3 per cent of the 16 developing utilities surveyed by the Asian Development Bank lost more than 15 per cent of their gross electricity generated in transmission and distribution; 37.5 per cent of them lost over 20 per cent. Losses can be broadly classified into technical and non-technical losses. All losses cause the utilities financial difficulties but technical losses represent economic losses for the countries. As it is expensive to reduce losses, the lowest possible losses are not necessarily the most economical. Optimum levels depend on the load and network characteristics of the system. Experts generally agree that the maximum tolerable losses should not be more than 15.5 per cent for a large power system with a reasonably complex network. The desirable limit is, however, much lower, and achievable, as demonstrated by some developing utilities that have losses below 10 per cent. Table 6 shows the experience of some developed economies of the world.

Table 5. Ranges of transmission and distribution losses

Losses (percentage)	Economies involved (percentage)
< 10.0	25.0
> 15.0	56.3
> 20.0	37.5

Note: Table 5 is based on an analysis of table 4.

Table 6. Transmission and distribution losses in selected developed economies of the world, 1990

	Percentages of gross generation
ESCAP region [a]	
Australia	5.7
Japan	4.2
Other region [b]	
Canada	7.5
France	6.3
Federal Republic of Germany (Former)	3.6
United Kingdom of Great Britain and Northern Ireland	7.5
United States of America	3.5
Union of Soviet Socialist Republic (Former)	8.1

[a] United Nations publication, Electric power in Asia and the Pacific, 1989 and 1990.

[b] *1992 Annual Bulletin of Electric Energy Statistics for Europe* (United Nations publication, Sales No. E/F/R.92.II.10)

It is becoming more and more difficult to secure investment funds for new capacity owing to competing investment needs against limited available funds. By reducing their system losses, utilities may ease their capacity addition problems. A reduction in the transmission and distribution loss rate from 20 to 15 per cent could in some cases be equivalent to one year's demand growth.

Conservation and co-generation

Major efforts are also needed in other areas of energy conservation and equipment efficiency improvement, such as the following: housekeeping and energy audit; economic system operation and control; equipment efficiency improvement; and reduction in fuel consumption by the generating plants. Reducing the fuel consumption of generating plants is the key to improving the overall efficiency of the system. Electric utilities in the ESCAP region are largely dependent on thermal power. Although most modern thermal plants can attain an efficiency of about 40 per cent, the operating conditions of the plants in the region are not so good on account of the poor manning and maintenance of traditional plants, achieving only 25-30 per cent thermal efficiency. For gas turbines, the figure is even lower. There are some exceptions, which show an efficiency range of 33-38 per cent.

An integrated total energy system (electricity and heat) could therefore greatly improve the overall efficiency in all energy-intensive industrial complexes. The first and simplest way is to construct more efficient plants in the expansion programme; the second, to retrofit or replace old, outdated plants or equipment; the third, to recover secondary energy from the waste heat (such as in combined cycle plants); and the fourth, co-generation (through combining heat and power production).

IV. CONCEPTIONAL CHANGES IN PLANNING

Until recently, the primary objective of load management had been either to reduce utility peak demand or to shift the consumption from the peak period to the off-peak period, with the ultimate aim of improving the load factor. In most developing economies this concept is still prevalent and applied. Latest thinking in this area adds to this load shaping objective the concept of demand-side management (DSM). In electricity planning a DSM programme is now recognized as a component resource.

Electricity is an unique commodity, which must be consumed as it is produced, with virtually no possibility of large-scale storage (apart from pumped storage hydro and batteries). Consequently, the entire generation, transmission and distribution chain must be harmonized to balance the supply and the demand at all times. This balance has also to be done in a situation where both the supply and demand are dynamic, with substantial variations over times.

Supply side

The power supply is characterized by many uncertain factors, some of which are predictable with a reasonable degree of confidence, whereas others are random. For example, scheduled maintenances are predictable but forced outages for various reasons are not. The easiest but not necessary the least cost way to balance the supply level with the demand is to have high reserve capacities of various types of plants. Power plants are normally grouped into three types: base load, intermediate load and peak load. Base load plants are large in size and use low-cost fuel such as run-of-the-river hydro, nuclear, coal or natural gas. These plants are operated at a more or less constant level throughout the day for 50-80 per cent of the year. Intermediate load plants are more moderate in size, and usually burn coal, oil or gas. Their outputs are changed throughout the day to follow the system load. Peak load plants are usually smallest in size and use costly fuel such as diesel, natural gas or hydro, running only at peak periods when the demand is very high. This accounts for the wide differences in electricity costs between peak and off-peak periods.

Demand side

Loads or demands are composites of a large number of customers using electricity for a variety of purposes such as heating, cooling, lighting and industrial machine drives. Loads vary according to the time of day, the month, the season and customers' choices or preferences. The end-use structure varies, too, according to the level of development of a particular economy. Utilities are now aware that the time has come to adopt a flexible management strategy to match the supply with the demand by managing both supply and demand sides.

V. LOAD MANAGEMENT TECHNOLOGY AND EXPERIENCE IN THE REGION

A. Demand side management technology

Various methods are applied for load management. The main ones are the tariff option, utility equipment control and end-use equipment control.

Tariff option

This option is often called the "soft option" or "indirect option". It is, however, the most important option, as it affects the other options. The others are generally combined with a tariff option because direct load controls use rate incentives and/or disincentives as a tool in motivating consumers to accept load management.

One major advantage of the tariff option is that it does not require capital investment and customers are free to consume as much electricity as they want. It is also an effective means of reducing load, as most large-scale customers are sensitive to price and will voluntarily minimize their loads.

Tariff policy is, however, a difficult issue for the utilities, particularly in developing countries, where the economic objective is often constrained by the financial, social and political objectives of the pricing policy. Therefore, in many instances, the tariff falls short of marginal cost.

The load shape of most utilities in the region is such that peaks are observed once or twice a day. This means that a high capacity must be kept up and maintained, which remains unutilized most of the time, therefore adding significant costs to the utility operation. A time of day (TOD) tariff system may be designed to charge a high price for these periods, while charging lower prices for the other periods. Since metering costs are high, only selected large customers with controllable loads are usual targets for TOD.

Utility equipment control

Voltage regulation, feeder control and power factor improvement are areas over which the utility can exercise direct control. Feeder control is considered an emergency measure only, involving the remote switching off and on of distribution feeders on an individual basis.

End-use equipment control

This category includes techniques that control air conditioning, water heating, pumps, heating, multiple loads and process loads. Many utilities with large commercial or industrial loads have targeted this sector for their initial load management programmes. They can thus obtain larger blocks of load control in a shorter period of time.

In high-income economies, major efforts have been made to control the load in the residential sector, which has the poorest load factor and constitutes the largest load sector. Except for a few economies in the region, however, residential load is not that significant. There is nevertheless a widespread tendency to improve the end-use efficiency of household appliances.

A number of load management systems are commercially available. These systems represent a wide range of technological approaches to demand-side load management, from simple mechanical, time-controlled devices to sophisticated computer control. Load management systems may also incorporate energy storage, co-generation, and small-scale generating equipment, such as solar and wind energy. The technologies may be broadly divided into the following three categories: local control and communication systems; thermal energy storage systems; and supplemental energy systems.[3] Table 7 lists the major categories of load management equipment and systems.

Table 7. Load management equipment and systems

I. *Load control and communication systems*

- Local controllers
  - Priority relays
  - Time controllers
  - Demand limiters
  - Load management thermostats
  - Programmable controllers
- Remote control systems
  - Ripple systems
  - Power line carrier systems
  - Radio systems
  - Telephone systems
  - Coaxial cable systems
  - Hybrid systems
- Communication information systems
  - Multi-building systems
  - SCADA

II. *Thermal energy storage systems*

- Ceramic heat storage
  - Room units
  - Central units
- Pressurized water heat storage
- Ice cool storage
- In-ground heat storage
- Combined heat and cool storage systems
  - Annual cycle energy systems
  - Daily cycle energy systems

III. *Supplemental energy system*

- Solar energy systems
  - Active systems
  - Passive systems
- Co-generation systems
- Dual heating systems

Source: Power System Load Management Technologies, Energy Department Paper No. 11 (Washington D.C., The World Bank, 1983).

---

3 *Power System Load Management Technologies,* Energy Department Paper No.11 (Washington D.C., The World Bank, 1983).

Load management technology is generally flexible and can be applied to a variety of circumstances. While utility conditions differ from country to country, the basic principles of the technology being essentially the same, developing countries of the ESCAP region can make the best use of the technology on the basis of customer considerations. The economics of the investment should, however, be considered before any decisions are made; although in many instances the option is more competitive than capacity additions.

Pricing, taxation, financial incentives and subsidies, education and propaganda are considered "soft" tools for demand management and are sometimes more useful in the medium or long run. Pricing is of particular importance in demand management in the electricity sector in maximizing the net economic benefits of consumption and in avoiding waste. With the rise in unit generation costs, the concept of marginal cost pricing is being accepted by many utilities. Although the application of long-run marginal cost in its stricter sense appears to be unrealistic for some developing countries, a two-stage compromise formula, suggested in a World Bank publication[4] could be considered. This would involve, first, the estimation of the long-run marginal cost to meet the objective of economically efficient pricing; and second, the adjustments to it to derive a realistic tariff schedule that satisfied constraints such as financial requirements, social-subsidy considerations, fairness, and metering and billing difficulties.

B. Demand-side management experience

Already in widespread use in the United States of America and in Europe, the concept of DSM is new in this region, particularly in developing economies. Some utilities (such as Malaysia, the Republic of Korea, Sri Lanka and Thailand) have introduced a time-of-use tariff for large industrial customers. Other measures such as off-peak load development and interruptible service are being tried in a few utilities. A pilot-scale DSM programme has been introduced in one or two utilities (including Thailand) in the region, while some others (including Indonesia, Malaysia) are studying the feasibility of introducing DSM.

The Asia Office of the International Institute for Energy Conservation (IIEC)[5] reports that the Thai DSM programmes can provide power for the Electricity

Generating Authority of Thailand (EGAT) at half the cost of building new generating capacity. Benefit-cost ratios for the DSM programmes ranged from 1.5 to 9.9. The DSM programmes will cover residential and commercial lighting, residential air-conditioners and refrigerators, new commercial building design and industrial motors. Table 8 shows the proposed budget and expected savings for the Thai DSM programmes. It is also reported that Thailand has been negotiating with the World Bank to get a grant of US$15.0 million from the Global Environment Facility to meet a part of the financing need.

Table 8. Budget and savings for Thailand DSM programmes

Programme	Budget (Millions of US$)	Estimated savings (Peak MW)
Lighting	101	133
Refrigerators	6	28
Air Conditioners	3	23
New commercial design and peak shaving	12	15
Industrial motors	19	30
Appliance testing	4	–
Consulting	4	–
Training	5	–
DSM administration	15	–
Marketing	9	–
Contingency	10.5	9
Total	188.5	238

Source: E-Notes, vol. III, No. 2, April - June 1993, Quarterly Newsletter of the International Institute for Energy Conservation.

VI. POTENTIAL ROLE OF DEMAND-SIDE MANAGEMENT AND OTHER DEMAND MANAGEMENT TECHNOLOGIES

Load management objectives

Energy conservation objectives can also be achieved sometimes through the use of load management techniques. A load management strategy must consider the characteristics of the utility, of the customers and of the load management technologies. Customer considerations relate to the type of load that may be controlled and to the institutional and business relationships between the customers and the utility. Utility considerations relate to the technical and economic need for load management and to the management and financial capability of the utility to support a load management programme.

---

[4] Mohan Munasinghe and Jeremy J. Warford, *Electricity Pricing: Theory and Case Studies* (published for the World Bank by the Johns Hopkins University Press, Baltimore and London, 1982).

[5] *E-Notes,* vol.III, No.2, April-June 1993, Quarterly Newsletter of the International Institute for Energy Conservation.

Load management can be performed either indirectly, by giving tariff incentives or disincentives to customers, or directly, through effective control of certain types of loads by the utility. To implement either or both of these methods, the following is required:

(a)   Development of an appropriate tariff structure;

(b)   Undertaking load survey and/or research;

(c)   Application or adoption of a technology suitable for the utility.

Load management potential

Any load management policy should be based on a careful study of the current status and future projections of demand and supply options to meet that demand. While supply-side planning is relatively easier, correct assessment of the future demand and its load shape is complex and may need sophisticated tools and methods for load research, depending on the size of a utility. The load management programme is therefore utility specific and may range from simple load control devices to sophisticated hardware- and software-driven systems.

Table 9 gives a breakdown of the sectoral electricity demand structure in selected developing economies of the Asian and Pacific region. Industry was the largest electricity-consuming sector, with a 64.2 per cent share, followed by the residential/commercial sector with 21.7 per cent share. Agriculture accounted for 8.6 per cent. The situation varied widely from economies with high industrial consumptions such as Fiji with 78.9 per cent, or China with 77.6 per cent, to economies with low industrial consumptions such as Hong Kong with 29.1 per cent, or Nepal with 34 per cent. In 75 per cent of the economies listed in the table, industrial electricity consumption was over 40 per cent. About 44 per cent had industrial consumption of over 50 per cent. Segregated data were unavailable, so residential and commercial sectors were grouped under one category. In some of the economies, this category also accounts for a good share of the electricity consumption

The industrial and commercial/residential sectors, therefore, are the obvious prime target groups for load management. The agriculture sector, even though it appears small, is also a good candidate for load management, as its load can be easily shifted from one period of the day to another, as needed.

The industrial sector comprises a diverse set of consumers. The degree of flexibility that industrial consumers have in absorbing power cut or load rescheduling varies widely. Industrial load management practice has normally been through tariff incentives. While that would still remain, new technologies are now available through which industry and the utility can work out a direct load control programme that would benefit both. Intelligent communications and metering devices are available to assist such programmes. The main advantage in initiating load management programme for industries is that their numbers are relatively small and they account for bulk consumption.

The potential for load control in the residential sector of some economies is also good. Large cities with a concentration of consumers having heating and air-conditioning loads are ideal for a load management programme. Household appliances such as refrigerators, freezers, cookers and washing machines consume a lot of energy. Residential loads can be controlled by local and remote control devices and through energy storage. Commercial loads could be either similar to residential or industrial loads depending on the type of commercial activities. Also in large cities, where shopping complexes, hotels and hospitals use large quantities of electricity, load management can be applied very successfully. In both the residential and commercial sectors, efficient lighting systems can help to significantly reduce demand without sacrificing the intensity or quality of lighting. Compared to other sectors, the current electricity consumption in agriculture is low in most developing countries of the region, although in some countries the growth of electricity consumption in the sector is high. The agricultural load pertaining to irrigation is a good potential area that deserves consideration for load management.

In determining the potential for load management, consideration should be given to the power system size, supply adequacy, load characteristics, load factor, generation mix, management capabilities and customer acceptance. These factors are country-specific and should be carefully weighed in the decision-making process both in terms of the viability and the technology.

Given the rapidly growing demand for electricity in developing economies of the region and the rising capital and environmental costs of meeting that demand, the benefits of DSM are considerable. Many countries have found that the DSM option is economically viable. It is therefore worth pursuing by utilities of the region that have the potential.

Table 9. Sectoral electricity demand in selected developing economies of the Asian
and Pacific region, 1990 (in GWh)

Economy	Total consumption	Residential/ commercial	Industry	Transport	Agriculture	Others
Bangladesh*	4 705 (100)	2 082 (44.3)	2 333 (49.6)			290 (6.2)
China (1989)	545 200 (100)	46 390 (8.5)	423 320 (77.6)	9 870 (1.8)	41 050 (7.5)	24 570 (4.5)
Fiji	422 (100)	89 (21.1)	333 (78.9)			
Hong Kong	23 834 (100)	16 908 (70.9)	6 926 (29.1)			
India	184 222 (100)	39 942 (21.7)	86 388 (46.9)	4 217 (2.3)	45 132 (24.5)	8 543 (4.6)
Indonesia	27 741 (100)	11 331 (40.8)	14 166 (51.1)			2 244 (8.1)
Republic of Korea	94 383 (100)	28 147 (29.8)	59 248 (62.8)	1 012 (1.1)		5 976 (6.3)
Malaysia	19 906 (100)	10 272 (51.6)	9 634 (48.4)			
Myanmar	1 840 (100)	859 (46.7)	933 (50.7)			48 (2.6)
Nepal	529 (100)	312 (46.7)	180 (34)		25 (4.7)	12 (2.3)
Pakistan	29 195 (100)	10 873 (37.2)	10 332 (35.4)	40 (0.1)	5 027 (17.2)	2 923 (10)
Philippines	20 651 (100)	9 589 (46.4)	9 763 (47.3)	29 (0.1)		1 270 (6.1)
Sri Lanka	2 608 (100)	1 023 (39.2)	910 (34.9)			675 (25.9)
Taiwan Province of China	82 668 (100)	23 219 (28.1)	49 191 (59.5)	434 (0.5)	1 611 (1.9)	8 213 (9.9)
Thailand	38 361 (100)	20 088 (52.4)	17 929 (46.7)		96 (0.3)	248 (0.6)
Viet Nam	6 382 (100)	2 701 (42.3)	2 985 (46.8)	73 (1.1)	623 (9.8)	
Regional Total	1 082 647 (100)	223 825 (20.7)	694 571 (64.2)	15 675 (1.4)	93 564 (8.6)	55 012 (5.1)

Source: Asian Development Bank, *Energy Indicators of Developing Member Countries of Asian Development Bank,* (Manila, Asian Development Bank, 1992).

Note: * Here, others includes non-energy uses.
Figures in parentheses show percentage of sectoral to the total electricity demands.

VII. CONCLUSIONS

Efficient utilization of electricity reduces environmental impact by delaying the construction of new generating plants. To make the power sector attractive for private capital investment, its efficiency has to be improved. Countries can seldom afford to lose energy. Optimum utilization of the existing system, optimum expansion planning and use of modern efficient technologies can curb the wasteful use of energy and electricity.

From the analysis of the situation in the region, there would appear to be good potential for controlling electricity demand through load management. DSM is an option open to utilities to manage their load. It would benefit not only the utility but also the consumers and the national economy.

DSM should be considered as a resource, as it reduces the electricity demand and thus a portion of the additional capacity requirement.

Despite its potential benefits, there are constraints that need to be addressed in adopting DSM systems. A thorough study of the system characteristics, including end-use load patterns, is essential for determining the suitability of DSM. Electricity pricing is a key factor in any efficiency improvement drive. Investment funding is required for hardware and software. Finally the motivation of consumers to accept the concept is extremely important.

The regional TCDC Working Group on Electric Power System Management provides countries in the region with an excellent opportunity for cooperation and for sharing their experiences in adopting new technologies such as DSM.

CHAPTER X

THE ESCAP ENERGY PROGRAMME

The global energy resource constraints perceived during the early 1950s and the early 1970s, have now, in the early to mide-1990s, been replaced by the constraints associated with the response to the global warming problem, which had been increasingly voiced since 1988. Even though, as late as 1990, at the time of invasion of Kuwait by Iraq, there was a renewal of concern over availability of oil supplies, it has been clear that in the energy sector the concerns on resource availability have been diminished. The currently advocated required response is the need to limit, and indeed to reduce, the demand growth for fossil fuels. This may be effected through pursuing the following: "no regrets policies", (1) decreasing the energy intensity, (2) fuel substitution and fuel switching from fossil fuels to non-fossil fuels, and (3) adoption of clean fuel technology.

Decreasing energy intensity may be achieved by

1. Increasing efficiency of energy production and energy consumption;

2. Energy conservation;

3. Promotion of energy efficient lifestyles.

Fuel substitution and fuel switching may be effected through:

1. Switching from coal to oil/gas and from oil to gas;

2. Switching power production from using fossil fuels to hydroelectricity, geothermal, solar and nuclear energy;

3. Switching from non-electrical energy to electricity, especially from internal combustion engines to electric motors.

Adoption of clean fuel technology (in respect of coal) involves the adoption of:

1. Pre-combustion technology (e.g. coal beneficiation),

2. Combustion technology (fluidized-bed, IGCC, etc.),

3. Post-combustion technology (pollution control and waste disposal);

and in respect of liquid fuels involves utilization of clean fuels (i.e. unleaded and low-sulphur gasoline).

The regional and subregional concerns in sustainable development pertain to the phenomenon of acid rain. Large consumption of fossil fuels, particularly coal, is associated with emissions of sulphur oxides (principally sulphur dioxide) and nitrogen oxides which combine with water to form sulphuric and nitrous(ic) acids. The acid rain affects and degrades soils and lakes. The required response is the incorporation of environmental protection equipment in energy production, in this case installation of sulphur (and nitrogen) removal equipment in power plants and industrial boilers, which may be pre- and/or post combustion.

The national or local concerns are those associated with urban centres (air pollution from industrial plants and from the transportation sector), and with sites for large power plants or dams.

Based on the foregoing, several "principal areas of activities" may be summarized within the subtheme of "energy development and management":

1. Improvements in management techniques, such as sectoral demand studies, analyses of conservation potential, optimum energy pricing, demand side management, etc. including questions on rural or equity imbalances;

2. Conservation and efficient utilization of energy;

3. Promotion of energy efficient lifestyles;

4. Fossil fuel substitution and fuel switching from coal to oil/gas and from oil to gas;

5. Promotion of the utilization of clean technologies and clean fuels;

6. Optimum utilization of electricity:

(i) Switching power production from using fossil fuels to hydroelectricity, geothermal, solar and nuclear energy;

(ii) Shifting from non-electrical energy to electricity, especially from internal combustion engines to electric motors.

7. Dissemination and promotion of new and renewable sources of energy (NRSE) technologies and penetration of such sources of energy;

8. Studies on the extent of and the required mitigation against acid rain;

9. A. Air pollution in urban/industrial centres

B. The siting of large power plants or dams.

Thus, the ESCAP subprogramme on environment and sustainable development under the subtheme "energy development and management" was prepared by making a judicious choice of activities from the above list. An examination of the list confirms that the principal areas of activities are to be found in several chapters of Agenda 21, the principal document of the United Conference on Environment and Development, which describes in summary form the programme to be undertaken by the whole world in order to address the urgent concerns on the environment. Among the most relevant chapters and programme areas are:

Chapter 4. Changing consumption patterns: *Developing methodologies, policies and strategies to encourage changes in unsustainable consumption patterns;*

Chapter 8. Integration of environment and development into decision-making: *Integrating environment and development at the policy, planning and management levels;*

Chapter 9. Protection of the atmosphere: *Promoting sustainable development: (1) Energy development, efficiency and consumption (2) transportation (3) industry;*

Chapter 14. Promoting sustainable agricultural and rural development: *Rural energy transition to enhance productivity.*

The following is a summary of the ESCAP work programme on the theme of energy development and management, as part of the subprogramme on environment and sustainable development.

2.1. Parliamentary services

(i) Parliamentary documentation

Reports to the Commission on the progress of the subprogramme

Reports of the Committee on Environment and Sustainable Development to the Commission

Reports to the Committee on Environment and Sustainable Development

(1) Energy scene and trends, including integration of environment in energy policy and planning

(2) Sectoral energy demand trends, the potential for interfuel substitution and energy conservation, and the status of demand-side management in Asia

(3) Prospects for enhancing energy efficiency in the Asian and Pacific region

(4) Operational efficiency improvements, including demand-side management of Asian and Pacific electric power utilities

(ii) Substantive services to:

(1) Commission sessions, 1994 and 1995

(2) Committee on Environment and Sustainable Development, 1994 and 1995

(3) Ministerial-level meeting on launching the regional space applications programme, 1994

(4) Ministerial-level conference on environment and development, 1995

(5) Inter-Agency Committee on Environment and Development

(6) Other committees and special bodies subsidiary to the Commission

2.2. Published materials

(i) Recurrent publications

(1) ESCAP Energy News (2 in 1994, 2 in 1995)

(2) Energy Resources Development Series (1 in 1995)

(3) Electric Power in Asia and the Pacific 1991 and 1992 (1 in 1994)

(ii)   Non-recurrent publications

    (1)   Environment and Development:

        (a) Environmental impact of air pollution on urban/industrial centres

    (2)   Energy Development and Management:

        (a) Energy efficiency guidebook for the Asian and Pacific region

        (b) New and renewable sources of energy supply and environmental management

        (c) Environmental management for power systems

    (3)   Manuals, Guidelines and Rosters: Environment and Sustainable Development

        (a) Roster of regional experts and institutions on conservation and efficient utilization of energy

(iii) Technical material

    (1) Update of the sectoral energy demand database and analysis of the energy situation in Asia

    (2) Update of energy planning software

2.3. Ad hoc expert group meeting

Energy resiliency and integration of environmental policy in energy development and management.

2.4. Operational activities

(i)   Advisory services

Energy development, conservation and management policy, environmental impact and risk assessment; energy resource options and technologies; new and renewable sources in energy mixes; rural energy supply; electricity demand management.

(ii)   Group training, seminars and workshops

Energy and efficiency; energy resource options and technologies; new and renewable sources of energy with emphasis on rural energy supply; electric power system management

2.5. Coordination, harmonization and liaison

The Committee on New and Renewable Sources of Energy and Energy for Development, of the Economic and Social Council, and regional working groups in various energy subsectors, the Department of Economic and Social Development of United Nations Headquarters, Food and Agriculture Organization of the United Nations (FAO), United Nations Industrial Development Organization (UNIDO), World Bank, International Atomic Energy Agency (IAEA), Asian Development Bank (ADB), Asian and Pacific Development Centre (APDC), Asia-Pacific Economic Cooperation (APEC), Association of South-East Asian Nations (ASEAN), East-West Center (EWC), International Institute for Applied Systems Analysis (IIASA), South Asian Association for Regional Cooperation (SAARC), World Energy Council (WEC), and other governmental, intergovernmental, and non-governmental organizations and their bodies, on matters related to energy.

ESCAP subprogramme 2, on environment and sustainable development, is funded through the regular budget of the United Nations. The programme would be supplemented by additional activities funded through extrabudgetary sources, i.e. multilateral agencies such as UNDP and bilateral donor countries contributing to ESCAP such as Australia, France, Japan, etc. These activities are mainly reflected as operational activities in the above summary. In the past, UNDP had been the major source of extrabudgetary funding: e.g. about $9 million for the Regional Energy Development Programme (REDP) for Asian developing countries implemented is in the period 1982-1992, and about $4 million for the Pacific Energy Development Programme (PEDP). For 1993-1997, UNDP has again designated ESCAP as the main executing agency for the Programme for Asian Cooperation on Energy and Environment (PACE-E) for the conduct of activities in six areas: energy and environment planning, coal development and utilization, natural gas and petroleum development, rural energy and environment planning, conservation and efficiency, and electric power systems management (see annex I) UNDP funding for PACE-E amounts to $3.5 million and will be supplemented by contributions from Australia ($1.8 million) and France ($ 800,000).

The UNDP-funded programme for the Pacific is being executed by the South Pacific Forum Secretariat.

ESCAP plans to implement the Asian component of the Global Energy Efficiency 2000 project, a global project proposed by the Economic Commission for Europe (ECE) to be implemented by all the United Nations regional commissions. In case the global project is not realized, the Asian component would be launched as Asia Energy Efficience 21 (see annex II).

The overall programme in succinct from is presented in annex III.

Annex I

ENERGY RESOURCES SECTION
ENVIRONMENT AND NATURAL RESOURCES MANAGEMENT DIVISION

ACTIVITIES PLANNED FOR JANUARY-JUNE, 1994

A. Activities of special significance

In addition to regular budget activities, ESCAP energy activities under the subprogramme environment and sustainable development would be boosted by the launching of the UNDP-funded operational project, Programme for Asian Cooperation on Energy and Environment. This $3.5 million project, with additional funding support from the Governments of Australia and France, would have six areas to be covered:

(1)   Programme element EEP: Energy and environment planning

(2)   Programme element CDU: Coal development and utilization

(3)   Programme element NG&PD: Natural gas and petroleum development

(4)   Programme element REED: Rural energy and environment development

(5)   Programme element C&E: Energy conservation and efficiency

(6)   Programme element EPSM: Electric power system management

One of the significant features of this programme is the emphasis on addressing the environmental dimensions of the issues in all areas of activities. Other important features include the active participation of the member countries in the implementation through the working groups and some of the national and regional institutions. In-kind contributions on a technical cooperation among developing countries (TCDC) basis is another strength of the project.

A brief summary of the activities that are planned to be carried out during the first six months of 1994 are given below.

Most of the activities would be initiated in early January 1994, though a few had already started in 1993. The programme, officially launched in October 1993, would continue through 1997. ESCAP has been designated as the main executing agency of the project.

B. A brief summary of activities planned for January-June, 1994

Activity EEP-1: Training in methodologies of sample design and techniques of data processing for improvement of sectoral energy demand.

The *immediate objective* of the activity is to strengthen capability of interested participating countries of the Working Group on Energy and Environment Planning in methodologies of sample design and techniques of data processing for improvement of sectoral energy demand assessments.

The activity would be carried out by the Asian Institute of Technology (AIT) in collaboration with ESCAP and the Working Group on Energy and Environment Planning. In June 1994, AIT will organize the first one-week regional training workshop.

Activity NG&PD-1: Strategic potential and implications of utilization of natural gas

The *objective* of the activity is to enhance the capability of the countries in making strategic choices among the different fuels for sustainable development and to increase the awareness of the strategic potential and environmental benefits of utilization of natural gas in the region.

ESCAP would prepare a report, updated and distributed after country visits in conjunction with consultants travel under NG&PD-2 started on 1 January 1994.

Country reports would be prepared by national focal points (updating the 1991 paper) with accomplishments, plans and environmental benefits/cost, regional links.

A regional workshop is scheduled to be held at ESCAP, in May-June 1994, to coordinate country plans and incorporate them in regional projects.

Activity NG&PD-2: Promotion of natural gas utilization in the energy market

The objective of the activity is to assist member countries in shifting to natural gas utilization as a more environmentally benign source of energy. Under this activity advisory services would be provided to the participating countries in development of appropriate strategies.

Activity CDU-2: Executive Seminar on Coal Technology and the Environment – Sydney and Hunter Valley – 1 week

The next seminar will be conducted in April 1994 along similar lines to those already held, though increasing emphasis will be given to environmental questions involved in the production and use of coal, reflecting the basic thrust of PACE-E. This activity has been subcontracted to the Joint Coal Board of Australia.

Activity REED-1: Evaluation of implementation of integrated rural energy environmental planning for sustainable rural development

The immediate objectives of the activity include:

(i)   To develop and disseminate operational knowledge on the concepts, institutions and technologies necessary for environmentally sound energy development in the rural areas;

(ii) To assist countries in the formulation of a programme of policies, strategies and activities for the implementation of an integrated approach to rural energy environmental planning for sustainable rural development.

During February-April 1994, a consultant would be engaged to undertake a review of work done on rural energy-environment planning and to prepare a synthesis paper for a regional workshop followed by a mission to China to review the experience of China and to assist national institutions in identifying links of national experience on rural energy planning with environment planning.

Activity C&E-1: National Strategies and regional cooperation in energy efficiency promotion

A workshop, scheduled for March 1994, is envisaged to constitute the start-up activity for the implementation of the new PACE-E conservation and efficiency work programme.

The main objective of this workshop is to assist the Governments in the region in formulating and implementing effective energy conservation promotion strategies. In addition, it would provide a regional forum for the discussion of substantive issues in the formulation of national energy efficiency promotion strategies; the workshop also aims at promoting collaborative initiatives among the participants from the various countries.

This activity will be implemented jointly by the Energy Management Centre of New Delhi, and ESCAP.

Activity C&E-2: Energy environment strategies for urban transportation

The implementing agency of this activity is the University of Hong Kong Center for Urban Planning and Environment Management.

The objectives are to:

(i)   Identify cost-effective strategies for short- and long-run reduction of negative environmental impacts and increase of energy efficiency in urban transportation;

(ii) Develop a network of officials who have policy-making responsibility in urban transportation.

A three-day workshop is scheduled to be held in the second quarter of 1994, following studies carried out in eight developing economies of the region.

Activity EPSM-1: Private sector participation in power generation and its consequences on environmental quality

The immediate objectives of the activity include:

(i)   To summarize the current and likely future state of private sector participation in power generation in developing Asia, especially opportunities and barriers to implementation.

(ii) To provide Governments with information on effective and practical means of ensuring that private electric power generation meets governmental environmental standards by sharing experience through TCDC and by analyzing the experience of other countries.

(iii)   To develop a TCDC network through which information and experiences concerning control of pollution from private power generation can be shared on a continuing, ongoing basis.

During the first half of 1994 a study outline for country studies and a resource paper would be prepared by an international consultant, followed by country studies to be prepared by five selected countries. These documents are to be reviewed at a meeting organized in a host country or at ESCAP in June 1994. A regional workshop is planned for later in the year.

Annex II

THE ASIA ENERGY EFFICIENCY 21 (AEE 21) PROJECT INITIATIVE : LEVERING PRIVATE SECTOR INVESTMENT IN ENERGY EFFICIENCY AND CONSERVATION

The availability and utilization of energy is as much a prerequisite as it is an integral part of economic expansion, modernization and growth. The developing countries in particular, including those of the ESCAP region, will continue to require more and additional energy resources, mainly fossil fuels, as they seek to achieve higher levels of industrialization and productivity, accelerated growth of gross national product and increased output, and to improve national and social welfare.

However, emissions resulting from the combustion of fossil fuels are a major source of environmental pollution of land, water and air, on a local scale (such as photochemical smog), on a regional scale, e.g. acid precipitation, and on the global scale, global warming.

The largest source of greenhouse gas emissions is combustion of fossil fuels, and therefore any emission reduction strategy should aim at least-cost energy approaches: fuel switching, improved energy conversion efficiencies and improved energy end-use efficiencies. This was recognized also in Agenda 21, the blueprint for sustainable development, adopted at the United Nations Conference on Environment and Development, Rio de Janeiro, 1992:

> "reducing the amount of energy and materials used per unit in the production of goods and services can contribute both to the alleviation of environmental stress and to greater economic and industrial productivity and competitiveness. Governments, in cooperation with industry, should therefore intensify efforts to use energy and resources in an economically efficient and environmentally sound manner."

Especially, improvements in energy end-use efficiency are generally regarded as being both technically feasible and economically viable, often having net economic benefits (i.e. savings) and relatively short pay-back periods. Apart from environmental reasons, Asian countries can take advantage of enhanced energy economy and efficiency in terms of reduced investments in energy supply infrastructure, reduced dependence on energy imports, reduced strains on the balance of payments, delayed depletion of scarce energy reserves and stimulation of indigenous technical competence and industrial capacity.

The existence of large and cost-effective opportunities for energy end-use conservation in buildings, transport, industries and businesses raises the question why such cost savings have not been realized. Throughout the Asian region, some barriers are common to saving energy, such as lack of information or awareness among consumers and institutional barriers.

A large number of investors and a majority of the consumers have insufficient information on available and rapidly evolving technologies and are uncertain regarding savings and cost-effectiveness, lacking capital and resisting buying equipment at greater purchase cost, requiring rapid payback. On the institutional and policy side, often significant subsidies to energy prices are found, especially in the case of electricity. Electricity tariffs are seldom time-differentiated. High import tariffs and non-fiscal barriers may stop energy-efficient products from entering a domestic market.

Annex III

Activities of ESCAP under subtheme of energy development and management as part of a coordinated plan of action on the implementation of Agenda 21 in Asia and the Pacific

Agenda 21 chapter & programme description	Agencies	Agencies' programme areas	Description of activities	Output	Budget 94-95	Collaborating agencies	Sources of fund	Linkage with other programme areas of Agenda 21	Geographical focus	Remarks
Chapter 7: Promoting sustainable human settlement development	ESCAP	Environment and sustainable development *Energy development and management*	1. PACE-E Conservation & Efficiency programme element C&E-2: Energy-environment strategies for urban transportation	W/s			UNDP	**Chapter 9** **Programme area 29: Promoting sustainable development:** **2) Transportation**		
Programme area 20: Promoting sustainable energy and transport systems in human settlements			2. Environmental impact of air pollution on urban/industrial centres	Publ.			XB			
Chapter 9: Protection of the Atmosphere	ESCAP	Environment and sustainable development *Energy development and management*	**Energy development** 1. PACE-E: Energy-Environment Planning programme element:					**Chapter 8** **Programme area 24: Integrating environment and development in decision-making**	Asia	
Programme area 29: Promoting sustainable development: (1) Energy development, efficiency and consumption			EEP-1 Training w/s on methodologies of sample design	W/s(3)	284,000	AIT	UNDP France			
			EEP-2 W/s on assessments to evaluate energy conservation potential							
			EEP-3 W/s on methodologies for integrating environmental considerations into energy planning and policy analysis							
			2. Energy resiliency and integration of environmental policy in energy development and management	EGM	34,900		RB			
			3. ESCAP Energy News	Publ.			RB			
			4. Energy Resources Development Series	Publ.			RB			
			5. PACE-E Natural gas & petroleum development programme element:							
			NGPD-1 W/s on potential for NG utilization	W/s			UNDP			
			NGPD-2 W/s on promotion of NG utilization	W/s			UNDP			
			6. PACE-E Coal development & utilization programme element:							
			CDU-2 TC on Coal technology & environmental management (2)	TC			Australia			
			CDU-3 Executive seminar on coal technology and the environment	Sem						
			CDU-4 Coal technology workshops (4)	W/s(4)						
			Electric power development 1. Electric Power in Asia and the Pacific	Publ.			RB			
			2. PACE-E Electric power system management programme element:							

Agenda 21 chapter & programme description	Agencies	Agencies' programme areas	Description of activities	Output	Budget 94-95	Collaborating agencies	Sources of fund	Linkage with other programme areas of Agenda 21	Geographical focus	Remarks
Chapter 4: Changing consumption patterns. **Programme area 7:** Developing national policies and strategies to encourage changes in unsustainable consumption patterns.	ESCAP	Environment and sustainable development *Energy development and management*	EPSM-1 W/s on private sector participation in power generation and its consequences for environmental quality **Efficiency** 1. PACE-E element: C&E-1 W/s on national strategies on energy conservation/efficiency C&E-3 W/s on energy efficiency standards and labelling C&E-4 W/s on the role of national energy conservation centres 2. Promotion of energy consulting services: Training workshops to promote consultancy in energy services (4 countries) 3. Energy efficiency guidebook for the Asian and Pacific region 4. Roster of regional experts and institutions on conservation and efficient utilization of energy	W/s & Publ. W/s(3) W/s(8) Publ. Publ.			UNDP UNDP Japan XB XB	 Chapter 9 Programme area 29		
Chapter 14: Promoting Sustainable Agricultural and Rural Development. **Programme area 55:** Rural energy transition to enhance productivity.	ESCAP	Environment and sustainable development *Energy development and management*	1. PACE-E Rural Energy-Environment Development programme element REED-1 W/s on integration of environment in rural energy planning 2. Wind energy 3. New and renewable sources of energy supply and environmental management	W/s TC(2) TCDC Publ.			UNDP Netherlands XB			

CHAPTER XI

PROGRESS AND FUNCTIONING OF THE REGIONAL ENERGY WORKING GROUPS[*]

A. Background and rational for regional cooperation

Developing countries of the ESCAP region are aware that energy sources are crucial to their economic development. Energy resources of the world are, however, unevenly distributed, and the ESCAP region is currently a net importer of energy. Even though energy production has been steadily increasing in the past decade or so the region's energy consumption has also been increasing at a rate commensurate with the dynamic economic developments.

Several subregions of the ESCAP region have excellent prospects for regional energy cooperation based on their energy resource endowments. The subregion of North-East Asia has vast resources of coal, oil and gas still to be tapped and used for the benefit of the region as a whole. The demand for energy is already very much present in the region: Japan and the Republic of Korea are dynamic economies with little of their own energy resources. The far-eastern part of the Russian Federation and the north-eastern and north-western parts of China are energy resource-rich areas with considerable potential resources waiting to be tapped. The subregion of South-East Asia also has considerable resources, although not as many as in North-East Asia, of natural gas and hydropower potential. The economies of this subregion are also experiencing dynamic and vibrant growth with prospects for further industrialization and development, thus requiring large inputs of energy. Subregional collaboration would enhance their collective security of energy supplies and increase the efficiency of their energy systems. Similarly, the subregion of South Asia could advantageously use the resources located within the subregion, such as natural gas in Bangladesh and the eastern part of India, hydropower potential in Bhutan, India, Nepal and Pakistan, and oil and gas resources in Afghanistan, the western part of India, the Islamic Republic of Iran, and Pakistan. The western part of the South Asian subregion could conceivably also be connected to the energy resource rich central Asian republics.

The type of subregional energy cooperation and collaboration envisaged above will not be realizable in the medium term. Such endeavours and efforts require very long lead-times and are usually a response to pressures exerted by economic realities, such as those pertaining in the European countries where such collaboration exits in all energy sectors. However, a beginning has to be made, and perhaps ESCAP could provide the forum to encourage discussion and dialogue on this topic.

One form of collaboration among developing countries is exemplified by the Organization of the Petroleum Exporting Countries (OPEC), formed by petroleum producing and exporting countries with specific aims and purposes relating to the area of petroleum exports. It was formed more than three decades ago. The oil crisis of 1973 displayed the strength of OPEC, so much so that in 1974 the member countries of the Organisation for Economic Cooperation and Development (OECD) formed its own organization, the International Energy Agency, to counterbalance OPEC.

Since then, both organizations have dealt with the overall energy problem, although each one had been established to deal with petroleum questions. The reason is, of course, that petroleum is an important and valuable commodity, and its use is principally as a source of energy, which is important for the economy. OPEC members are from developing countries only, even though some developed countries has been petroleum exporting countries before 1960 and since the 1970s some other developed countries have become petroleum exporting countries; IEA consists of members only of developed countries, most of them oil importing countries; developing countries are not eligible for membership, even though many developing countries are also oil importing countries. (The Republic of Korea may soon become a member).

Clearly both institutions have their *raisons d'etre*. The members recognize the need for the institution and support and sustain its existence. It is equally clear that both had been established to deal directly with questions of supply and demand for petroleum and petroleum products. But both institutions are also engaged in the conduct of studies, pertaining to not only the specific area of petroleum but also the overall energy problem, as well as area closely related with energy.

[*] Note by the ESCAP secretariat for the Expert Group Meeting Preparatory to the First Session of the Committee on Environment and Sustainable Development, Bangkok, 30 September - 2 October 1993 (No. NR/PCESD/8).

Another example of an organization serving the interests of developing member countries, in this case in Latin America, is OLADE, the Latin American Energy Organization. OLADE is governed by a board, consisting of the ministers of energy of its member countries, and its secretariat is supported by contributions from its members.

For Asia and the Pacific, there is no question about the wide scope for regional cooperation in energy. Despite the large disparities in stage of economic development, the great variety in natural resource and energy endowments, and the stark differences in human resources, developing countries of the region could fruitfully collaborate in exchanges of information, assist each other in meeting training needs, and benefit from transfer of know-how and experience, in energy development and management, in areas such as energy planning, the coordination of policies, the energy trade such as in petroleum products, in gas and electricity networks, and also in environmental mitigation measures. Such collaboration could later develop into regional energy cooperation. However, developing countries of the region are constrained by lack of foreign exchange and lack of funds in general to undertake a sustained effort of regional cooperation in energy. In the Asian and Pacific region countries still require a source of grant funds to enable them to participate in a series of meetings, as would be called for in a regional cooperation programme covering the energy sector.

Thus in 1988, after more than five years of implementation of two regional programmes in energy, namely the Biomass, Solar and Wind Energy Programme, funded by the Government of Japan with supplementary assistance from the Government of Australia, and the Regional Energy Development Programme (REDP), funded by the United Nations Development Programme (UNDP) with supplementary funding from the Governments of Australia and France, it was decided in two separate ESCAP forums to establish networks of regional energy working groups. The first was the network of the regional working groups on new and renewable sources of energy, with each working group dealing with only one specific area of new and renewable energy such as biomass, solar, geothermal and wind, and the second was the network of six regional energy working groups within the framework of REDP, each dealing with one specific topic or energy type: natural gas, coal, energy planning, energy conservation, electric power, or rural energy planning. It was felt too premature to establish a full-fledged energy organization, as many countries were unable to contribute the funds necessary to support such an institution. The countries intimated that grants, such

as from UNDP and bilateral sources such as Australia, France, Germany, Japan, the Netherlands and Norway, should be available to enable them to develop cooperative programmes.

B. Establishment by ESCAP and UNDP of regional energy working groups

1. Working Groups under the REDP framework

The third session of the Tripartite Review Conference of the Regional Energy Development Programme, held in Kuala Lumpur in August 1989, approved the work programme for phase IV of REDP for implementation during the period 1990-1991. The work programme was divided into six subprogrammes: energy planning, natural gas development, coal development, energy conservation, electric power development, and rural energy planning and development. Each of the subprogrammes comprised several activities, of which at least one in each subprogramme was to conduct a regional seminar or workshop. During the conduct of the seminar or workshop, normally of three-to-five days duration, a special session was devoted to discussion in working groups of topics or questions of interest to member countries. The seminar or workshop participants acted as working group members representing their respective countries.

The working groups had the following aims and immediate objectives:

(a) To promote and develop a cooperative arrangement for regional cooperation to strengthen the national capabilities of the participating member countries in a specific area of energy;

(b) To promote and operationalize technical and/or economic cooperation programmes among participating member countries through (or as a result of) implementation of working group activities.

Six working groups within the framework of REDP were established during the period 1990-1991. The Working Group on Rural Energy Planning and Development met during the Executive Seminar on Rural Energy Planning and Development, held in China and Thailand, in October and November 1990. The State Planning Commission of China acted as the coordinator of the Working Group. The Working Group designated China, the Philippines and Sri Lanka to form a subgroup entrusted with the task of developing project profiles based on the programme outlines agreed by the Working Group. For this purpose a meeting was organized on a TCDC (technical cooperation among developing countries) basis in Beijing by the State Planning Commission. The draft project profiles were circulated among members and then finalized and submitted by China to the Tripartite Review Conference of the Regional Energy Development Programme, at its fifth session held in Kuala Lumpur in August 1991.

A similar procedure was followed by the other working groups. The Working Group on Natural Gas Development met during the Executive Seminar on Natural Gas Development, held in Indonesia and New Zealand in February 1991. The designated coordinator was the State Gas Corporation of Indonesia, whose President Director presented the proposed project profiles on natural gas to the Tripartite Review Conference at its fifth session. The Working Group on Energy Conservation discussed a cooperative programme for the period 1992-1996 during the Regional Training Workshop on Energy Conservation in Small and Medium-scale Industries, held in New Delhi in March 1991. The Energy Management Centre of the Department of Power under the Ministry of Energy, India, is the Coordinator of the Working Group.

The fourth working group formed was the Working Group on Coal Development, which discussed a future programme in conjunction with an executive seminar held in Australia. The Joint Coal Board of Australia acted as the coordinator of the Working Group and its representative submitted a proposal to the Tripartite Review Conference as its fifth session. Subsequently, the Working Group on Electric Power Development was convened during the Regional Workshop on Economic Load Dispatching and Demand Management in Electric Power System Management, held from 29 July to 2 August 1991 in Malaysia. The Department of Electricity Supply of the Ministry of Energy, Telecommunications and Posts of Malaysia is the coordinator of this working group. The sixth and last working group was the Working Group on Energy Planning, which was formed during the fifth session of the REDP Tripartite Review Conference. The REDP Senior Coordinator assumed the responsibility as coordinator of the working group and finalized the project profiles based on the work programme outline agreed upon by the working group.

The establishment of the six regional working groups has proceeded smoothly. The response of the countries of the region has been very encouraging. For example, 17 countries have indicated their interest in participating in the Working Group on Energy Conservation and 15 countries are participants in the Working Group on Electric Power Development. Technical support for all working groups has been provided by the ESCAP secretariat through REDP and the Natural Resources Division. The Division of Industry, Human Settlements and Environment has also provided support to the Working Group on Energy Conservation. Additional support has also been provided by the Division of Statistics and the Development Planning Division.

2. Regional working groups on new and renewable sources of energy

The formation of the working groups on new and renewable sources of energy followed a procedure analogous to the REDP working groups, with working group discussions conducted during a session of the regional seminar or workshop on a particular topic of new and renewable sources of energy. The aims and objectives of the working groups were also similar. However, there were differences. The principal activity was not funded directly by UNDP through ESCAP but by a bilateral donor country, or by UNDP through another agency. There was no equivalent of the REDP Tripartite Review Conference to review and endorse the proposals from the working group. Unlike REDP, which was established for the benefit of Asian member countries (because Pacific island countries at that time obtained UNDP grants through the Pacific Energy Development Programme (PEDP)), activities were open to all member countries of the ESCAP region.

The first working group formed was the Working Group on Wind Energy Development and Utilization, effected in conjunction with the organization of the Regional Expert Group Meeting on Wind Energy Technology in China, held in November-December 1990. China agreed to host the secretariat of this working group. The second working group to be established was the Regional Working Group on Geothermal Energy Development and Utilization, which was set up during the International Workshop on Geothermal Energy Training, held in the Philippines in November 1991. Both working groups succeeded in formulating work programme outlines for their future work.

The establishment of other working groups is being considered. Among these are a regional working group on solar photovoltaic technology, a regional working group on solar thermal energy development and utilization and a regional working group on biogas production and utilization.

3. Regional cooperation by means of technical cooperation among developing countries

The type of cooperation effected through the working group mechanism is known as technical cooperation among developing countries (TCDC). This terminology is applied to a particular form of cooperation, namely technical exchange through mutual visits with local travel and subsistence costs being met by the host country and international costs provided by an international agency. This type of cooperation has been conducted by ESCAP, with supplementary funds from bilateral donor countries or from UNDP.

Such TCDC cooperation through ESCAP is already being effected in the area of new and renewable sources of energy. The three-country new and renewable sources of energy subprogramme of REDP has been successfully implemented to the satisfaction of all three participating countries, China, the Democratic People's Republic of Korea, and Mongolia. The Liaoning Province Energy Research Institute of China is the coordinator. The subprogramme has been mutually beneficial in that exchanges of information and experience have been effected on various new and renewable sources of energy technologies and applications in countries with similar climates and agro-systems. Another series of activities pertained specifically to solar photovoltaic systems for remote villages. The donor country supplying the solar photovoltaic systems was Thailand and the beneficiary country was Myanmar. Similar exchanges have recently been implemented in wind energy technology between China and Sri Lanka, and between China and Thailand, and also in charcoal production between China and Vanuatu.

In the future such exchanges should be initiated and discussed during meeting on project activities being undertaken by ESCAP.

C. Functioning of the regional energy working groups

The experience of the implementation of energy activities during the period 1990-1992 in respect of the regional energy working groups has been very satisfactory. The networks have been established and each of the working groups have succeeded in formulating work programme proposals for the future (in the case of the REDP working groups, for the period 1992-1996, in conformity with the period of the fifth programming cycle). Leadership has been provided by the coordinating country through the host of the secretariats of the working groups. The host countries have also contributed funds for the organization of the main activities as well as for the working groups.

The foundations for a potential blossoming of regional cooperation in energy have thus been laid down. However, there remains much work to be done. The countries of the region that are already at a more advanced stage of development are eager to share their experience with other countries at a lower stage of development. Countries at the beginning of their economic reform programmes are enthusiastic about applying the analytical methods and methodologies commonly used for market economies. Old challengers remain, such as the vulnerability of oil importing countries to oil price fluctuations or the rationalization of tariff structures, and new challenges have emerged, such as the incorporation of environmental considerations in energy planning. Regional cooperation in energy effected through the regional cooperative mechanisms reported in this note provides the necessary base on which to meet these challenges.

UNDP has now agreed to launch with ESCAP a new programme on energy and the environment: Programme for Asian Cooperation on Energy and the Environment (PACE-E). The programme will make full use of the working group "infrastructure" already built, with some modifications: (a) the participating will be requested to form "national focal groups" on energy and environment, instead of simply designating a "national focal point" to act as the national counterpart for the programme; and (b) the names of the working groups have been reformulated in the project document to reflect the energy-environment interrelationship.

The launching of the PACE-E activities, probably in September 1993, will coincide with the convening of this Expert Group Meeting. It is suggested that the experts discuss the possibilities for regional cooperation that could conveniently be realized in the coming years. It may be desirable to have a meeting of senior experts at policy level to discuss a possible future cooperating and coordinating mechanism to oversee and guide the activities of all working groups in the field of energy: thus all areas of energy development and management could be coordinated by the same forum without any duplication.

The secretariat invites the experts to deliberate these issues and to make recommendations to the Committee on Environment and Sustainable Development accordingly. The principal activities envisaged are coordination of the energy development and management activities under the Subprogramme on Environment and Sustainable Development, execution and implementation of PACE-E, consideration of Asia Energy Efficiency 21, and TCDC activities in selected areas not covered by the above.

CHAPTER XII

IMPLEMENTING UNCED:
A PROGRESS REPORT FROM THE EUROPEAN COMMUNITIES
TO THE UN COMMISSION ON SUSTAINABLE DEVELOPMENT*

---

*   Contribution to the work of the Committee on Environment and Sustainable Development

1. OVERVIEW

1992 was a watershed for the promotion of sustainable development by the European Community. During the year since Rio, the European Community and its Member States have taken a series of important decisions which lay the foundations for sustainable development policies, both at home and abroad. The Treaty on European Union, signed at Maastricht in February 1992, marks the next stage in the evolving process of European integration and makes the pursuit of "sustainable growth respecting the environment" a central objective of Community policy-making; the Treaty also reinforces the constitutional basis for the Community's environment and development policies.

At UNCED itself, the Community made a substantial contribution to the successful outcome of the "Earth Summit". The Community subscribed to the Rio Declaration on Environment and Development, the Agenda 21 and the Statement of Forest Principles and signed the Biodiversity and Climate Change Conventions. The Community also took specific actions at Rio to demonstrate its long-term commitment to push forward international sustainable development efforts, including the announcement by the Community and its Member States of a special initiative to make available 3 billion ECUs[1], including new and additional resources, for implementing Agenda 21 in developing countries (see section 5) and the confirmation of its decision to stabilise carbon dioxide emissions at 1990 levels by 2000, on the basis set out in the Joint Council Conclusions of October 1990. Soon after Rio, Community Heads of State and Government meeting in Lisbon agreed an 8-point plan to implement the Agreements and Conventions.

The strategic environment and development policy frameworks for the follow-up to UNCED were then set at meetings of the Development and Environment Ministers. In November, the Development Council laid down priorities for EC 'Development Co-operation Policy in the run-up to 2000', agreed that the Community and its Member States would seek to provide an initial tranche of 600 million ECU in the first year of the implementation of the Community initiative, and adopted a series of measures to promote clean and efficient energy use in developing countries. In December, the Environment Council adopted a resolution on the new Community programme of policy

_____
[1] 1 ECU = 1.24 US dollars

and action in relation to the environment and sustainable development, "Towards Sustainability", and concluded that this programme was "an appropriate point of departure for the implementation of Agenda 21 by the Community and the Member States". The Council also updated and reinforced the 8-point Lisbon plan. In May 1993, the Development Council reached the conclusions which are annexed to this report.

The Commission's work programme for 1993-94 stresses the importance of pursuing the results of Rio by integrating environmental concerns into the definition and implementation of Community policies in other areas, and detailing the Community's financial commitment to developing countries. The Commission has also initiated a process to develop the necessary internal institutional mechanisms and procedures to make sustainable development a permanent and integral part of Community policy-making.

Since Rio, the Community has participated actively in the wide array of international activities to implement UNCED, notably the 47th session of the United Nations General Assembly, the follow-up meetings for the Climate and Biodiversity Conventions, the preparations for a Desertification Convention and the Conferences on Straddling and Highly Migratory Fish Stocks and on the sustainable development of small island developing States. The Community is aiming to ratify and publish national strategies for the Climate Change and Biodiversity Conventions by the end of 1993; the Environment Council has already given its agreement in principle to Community ratification of the Climate Change convention. The Community is preparing a number of actions to take forward the non-binding Forest Statement adopted by UNCED, both internally and at the international level.

By the end of 1993, the Community will have completed the first stage of implementing its UNCED commitments. Within the Community, the process of integrating the Rio Declaration, Agenda 21 and the Statement of Forest Principles into existing Community policies should be well underway. The implementation of the new initiatives to assist developing countries in achieving sustainable development and to promote transfer of environmental technology should have started. Both the Climate Change and Biodiversity Conventions should have been ratified, and agreement should have been reached on Community, as well as national, strategies to implement them. The Community experts all other developed and developing countries to give effect to their own commitments at Rio.

2. RIO DECLARATION AND AGENDA 21

2.1. Internal Community Policies

At its December 1992 meeting, the Environment Council agreed that the new European Community Programme of policy and action in relation to the environment and sustainable development ("Towards Sustainability") provided "a comprehensive as well as a strategic approach to sustainable development", whose implementation will make "a major contribution to the follow-up to Agenda 21 by the European Community and its Member States". The programme was drawn up during the preparations for UNCED, and thus shares many of the principles contained in the Rio Declaration and the objectives of the Agenda 21 action programme. The new programme is based explicitly on the polluter pays, the prevention and precautionary principles, as well as on the imperatives of integrating environment and development policies, addressing unsustainable consumption and production patterns, building partnerships and sharing responsibility among state, economic and civil actors and promoting international action for sustainable development.

For the Community, dialogue and partnership between governments, economic actors and civil society at all levels is an essential precondition for sustainable development. The Community has adopted a Directive to ensure full public access to environmental information, and has taken a number of measures to make its decision-making processes more transparent to its citizens (including the televising of ministerial sessions). Representatives of the European Parliament and European environment and development non-governmental organizations (NGOs) were part of the Community delegation in Rio. The Community intends to deepen its working relationship with the NGO sector in the implementation of the UNCED Agreements. Within the Community, a new Consultative Forum representing business, labour unions, local government and environmental organizations will be established in 1993, in line with Agenda 21 recommendations. As part of its co-operation activities with developing countries, the Community provides financial assistance through NGOs, and places particular emphasis on upgrading the role of women.

To explore further the positive linkages between a healthy economy and a healthy environment, the Commission will host a major public conference on the links between environment, employment and "green accounting" in the latter part of 1993, in collaboration with leading European environmental groups.

For the rest of the 1990s, the new programme of policy and action in relation to the environment and sustainable development will be the principal instrument for achieving sustainable development within the Community. The programme lists priority environmental challenges for the Community and sets out policy objectives. It has also selected some target sectors in the economy which have a crucial bearing on the transition to sustainability, and has called for a broadening of the range of policy instruments to include economic measures and voluntary agreements. A system for monitoring Community performance against the programme's objectives is being developed, and the Commission will publish regular progress reports. The programme will be reviewed fully and upgraded at the end of 1995. The Community's record in implementing its UNCED commitments will be contained both in the regular progress reports and in the 1995 review.

While the integration of environmental factors is not a new goal for the Community, the agreements reached at UNCED and the "Towards Sustainability" programme have added a new urgency to this imperative. Since Rio, the Community has taken a number of policy initiatives which demonstrate that environmental integration is becoming a reality, including:

– *Regional Development*

In February 1993, the Commission published proposals for the management of the Community's Structural Funds for regional development during the next five years (1994-99). The proposals include a number of new features that will go some way towards integrating environmental considerations.

The Council has also decided to establish a new Cohesion Fund to provide financial assistance for environmental and transport infrastructure projects in four Member States (Greece, Spain, Portugal and Ireland).

– *Agriculture and Forestry*

In June 1992, as part of the reform of the Common Agricultural Policy, the Community adopted two regulations to stimulate more sustainable agricultural practices. The first is intended to promote ways of using agricultural land which are compatible with the protection and improvement of the environment and biological diversity. The second regulation introduces a Community scheme to encourage afforestation as an alternative use for agricultural land

and to develop forestry activities in farms. Finally, in July 1992, two regulations were adopted strengthening the Community's action for the protection of forests against atmospheric pollution and fires.

– *Transport*

Following the Corunna and Shetland tanker accidents, a joint Environment and Transport Council underlined the need expressed in Agenda 21 for integrated coastal zone management and concluded that "concerted action at both national and Community levels and within the framework of the International Maritime Organization is essential to reduce to a minimum the dangers posed to the environment by the transport of oil and other dangerous cargoes". The Commission has subsequently produced a strategy paper on "Safe Seas".

In December 1992, the Commission also issued a new White Paper on the "Future Development of the Common Transport Policy" which provides a global approach to tacking Europe's transport problems in the coming years. The paper builds on an earlier Green Paper "A Community Strategy for Sustainable Mobility".

– *Industry*

In late 1992, the Council of Industry Ministers adopted a Resolution on industrial competitiveness and environmental protection, which recognized that a high level of environmental protection has become not only a policy objective of its own, but also a precondition for industrial expansion.

The Community has also pioneered the application of a new range of policy instruments to change current unsustainable consumption and production patterns. In March 1992, the Environment Council adopted a regulation to introduce a voluntary Community "eco-label" system to assist consumers in making informed choices between different products; the first eco-labels will appear on an initial range of products during 1993.

In March 1993, Community Environment Ministers agreed to launch a second voluntary initiative for an environmental management and auditing system for industrial sites ("eco-audit"). Participating companies will be required to establish and implement "environmental policies, programmes and systems, undertake systematic, objective and periodic evaluations of their performance and publish annual environmental statements". The Commission has also proposed that there should be a Community debate on liability for environmental damage.

2.2 Regional Cooperation

The EC Commission participated in the Conference of Environment Ministers from the UN-ECE Region which took place in Lucerne (Switzerland) on 28-29 April 1993. An Environmental Action Programme for Central and Eastern Europe (EAP) was endorsed by the Ministers. This Programme represents a broad consensus on a general approach based on three main elements:

– integration of environmental considerations into the process of economic restructuring, to ensure sustainable development;

– immediate assistance programmes;

– institutional capacity building, including effective legal and administrative frameworks.

The EC is already playing an important role in these areas. According to an official report circulated at the Conference, 49 per cent of all funding for Central and Eastern European countries (CEECs) has been channelled through the EC PHARE and TACIS programmes.

In June 1993, the Community will participate in the Second Ministerial Conference on the Protection of Forests in Europe which is expected to adopt general guidelines for the sustainable management of European forests, as well as resolutions concerning the conservation of biological diversity in forests, co-operation with Central and Eastern European countries in the forestry sector and the protection of forests against the potential effects of climate change.

2.3. International Community Policies

– Ozone Depletion: The European Community and its Member States strongly support the second amendment to the Montreal Protocol agreed at Copenhagen in November 1992. At the December 1992 Environment Council, Member States agreed not only to implement the Protocol's accelerated schedules to reduce the production and consumption of ozone depleting substances, but exercised their right to go further by phasing-out chlorofluorocarbons and carbon tetrachloride by December 1995, rather than December 1996 as required under the Protocol. The Commission is preparing proposals to take forward the Copenhagen agreements on transitional substances (HCFCs) and methyl bromide.

– Hazardous Wastes: The Community has long taken a leading role in controlling the international movements of hazardous wastes. In 1989, as part of the Lome IV Convention, the Community decided to ban the export of hazardous and radioactive wastes to the 69 African, Caribbean and Pacific countries. The Community and its Member States are signatories of the 1989 Basel Convention. Agreement was reached in the March 1993 Environment Council that ratification of the Convention by the Community would occur by February 1994 at the latest. The provisions of the Convention will be implemented under a new Council Regulation adopted in February 1993 which will be applicable from May 1994. This Regulation bans all exports of hazardous wastes for disposal except to EFTA countries, which are also party to the Basel Convention, and introduces new restrictions on shipments of hazardous wastes for recovery. In March, the Council asked the Commission to examine the circumstances which justified the call for a total ban on hazardous waste exports to developing countries and to submit appropriate proposals to the Council at the earliest possible date.

– Desertification: The Community and Member States have already made significant efforts to control the threat of desertification, particularly in Africa, contributing over 1 billion ECU in assistance to date. Some initiatives aimed at combating desertification in the South of Europe have also been undertaken.

In the light of its past record, the Community and member States will participate positively in the work of the Intergovernmental Negotiating Committee on Desertification.

3. OTHER ASPECTS OF UNCED FOLLOW-UP

3.1 Climate Convention

In early June 1992, the European Commission published a strategy to support the Council's agreement to take action with the aim of stabilising carbon dioxide (CO_2) emissions at 1990 levels by the year 2000 (which assumes that other leading industrialized countries will do likewise): this strategy included proposals to monitor carbon dioxide emissions, promote energy efficiency (SAVE), encourage alternative energy sources (ALTENER) and to levy a joint tax on energy inputs and carbon emissions. Later the same month, the European

Community and its Member States signed the Climate Change Convention in Rio, and made a declaration calling for the prompt start to the implementation of the Convention.

In March 1993, the Environment Council of Ministers decided to take the necessary steps for the Community and the Member States to ratify the Convention as soon as possible and at the latest by the end of 1993. Agreement was also reached on the introduction of a common monitoring mechanism for CO_2 emissions, whereby Member States will draft, publish and implement national strategies to control CO_2 emissions. Discussions are continuing on the remaining elements of the CO_2 stabilisation strategy, particularly concerning the carbon tax proposal.

3.2 Biodiversity Convention

In June 1992, the Community signed the Biodiversity Convention in Rio. In December 1992 the Environment Council agreed to establish the basis for ratification of the Biodiversity Convention with the aim of achieving ratification by the end of 1993 and to prepare national strategies by the same time. Preparatory work is progressing well, and the year end deadline for ratification is likely to be met.

Within the Community, the Convention will be implemented in particular by means of the recent Habitat Directive, which was approved by the Council of Ministers one month before UNCED in May 1992. The Habitat Directive aims at the maintenance of European biodiversity through sustainable land management practices, and the establishment of the Nature 2000 network of protected sites. Other directives like those on environmental impact assessment and on the contained use and the deliberate release into the environment of genetically modified micro-organisms will also play an important role in the implementation of the Convention.

3.3 Statement of Forest Principles

The Community and its member States regard the non-binding statement of forest principles adopted at UNCED as a first step towards the preparation of a possible Global Forest Convention. The Community supports the review of the implementation of the forest principles under the aegis of the UN Commission on Sustainable Development and all further efforts towards the preparation of a possible Global Forest Convention. At the Lisbon European Council, in June 1992, the Heads of State and Government made the commitment to publish national plans for the implementation of the Statement of Forest Principles.

To give effect to the Statement of Forest Principles within the EC, the Commission is preparing a strategy on the management, conservation and sustainable development of the Community's forests. This strategy will be supplements by a list of priority areas for additional action by the Community in the forestry sector and a common format for Member States to report on their progress in implementing the Forest Statement.

The Community is also committed to the long-term support of developing countries' efforts to promote sustainable forestry policies and practices. In 1993 the Community will allocate more than 150 million ECUs for actions related to the protection and rational management of tropical forests.

4.   TRANSFER OF TECHNOLOGY

I   Introduction

Chapter 34 of Agenda 21 stresses the vital role of environmentally sound technologies in supporting the prudent management of the environment and development. The objectives can be described as follows:

a)   To promote the access to and the transfer of environmentally sound technologies without neglecting the development of indigenous ones;

b)   Human resource development;

c)   Strengthening of institutional capacity to carry out research and to conceive and execute sustainable development strategies;

d)   Integrated sectoral assessments of technology needs, in accordance with countries' plans, objectives and priorities as foreseen in the implementation of Agenda 21 at the national level;

e)   to promote long-term technological partnership between holders of environmentally sound technologies and potential users.

II.   Community approach and activities

In their announcement on finance, the Community and its Member States defined, inter alia, technology transfer as a cross-sectoral priority for projects and programmes to be financed in key Agenda 21 sectors in the context of their financial initiatives. In the follow-up to Rio, the Community has focused on several key Agenda 21 sectors with a view of identifying actions to be initiated by the Community and contribute to the objectives of Chapter 34.

Strengthening Commercial Links

Beside the traditional instruments for cooperation and development, the Commission of the EC is progressively developing a more and more detailed framework of *economic cooperation.*

Economic cooperation is designed to serve the *mutual interests* of the European Community and its partner countries.

Its purpose is to strengthen the productive and institutional capacity of the partner country so as to make its economic environment more receptive to investment and trade, whilst at the same time mobilising the resources (such as capital, technology, know-how, marketing and managerial capabilities) of European economic operators and facilitating the transfer of those resources, under market conditions, to local operators.

Economic cooperation is particularly effective in countries which are implementing open-door policies and which have the potential for rapid growth. It encourages direct involvement of the private sector, as a driving force behind economic expansion, and is in line with the trends emerging in a number of developing countries: the tendency for the private sector to play an increased role; the introduction of market rules and mechanisms; the orientation of economies toward international trade; the modernisation and expansion of infrastructure; the more effective use of human resources in education and research; the increasing propensity of people to save, etc.

In order to facilitate economic cooperation, the Community is implementing different types of activities with its partner countries involving directly the private sector. These include, for example, industry-specific technical assistance programmes, the setting up of Business Information Centres and multi-sectoral actions such as programmes in standars and patents. The majority of activities involve transfer of technology between the private sectors of the EC and its partner countries.

Investment promotion is also of particular importance in the context of economic cooperation. For this purpose, the Commission has created a specific instrument of investment promotion to facilitate joint ventures between local operators and their European counterparts – the ECIP facility (European Community Investment Partners).

This focuses particularly on joint-ventures or licensing (new form of technology transfer). These operations constitute, among others, a very efficient vehicle of technology transfer, one of the main objectives of the instrument. Its budget for 1993 is 40 million ECU. It is targeted on the Asia, Latin American and Mediterranean countries.

C.   Clean and Efficient Energy for Development

In November 1992 the Council of the European Community adopted guidelines for the transfer of clean and efficient energy technologies for sustainable development.

It was emphasized that an essential objective of cooperation with all developing countries in the field of energy is to contribute to the promotion of sustainable energy production and consumption through the implementation of effective energy policies and the introduction of more efficient technologies for sustainable development.

Special attention will be paid to :

–   the progressive substitution of high carbon intensity fossil fuels by lower CO_2 emission conventional fossil fuels, and the development of clean combustion techniques;

–   the development of renewable energy resources.

In order to achieve the objective the European Community intends to:

a)   strengthen its technical assistance activities with emphasis on institutional capacity building, policy planning, pricing techniques, management and the improvement of technological capacity

b)   promote through instruments of economic cooperation, joint ventures and other investments which will support clean technologies. The European Investment Bank and the EC Investment Partners are expected to play the major roles

c)   adapt and extend existing programmes of scientific co-operation and promotion of new energy technologies in order to include in their scope the needs of developing countries and contribute to their technological progress.

The EC Commission services are currently working on proposals for the implementation of these guidelines which could be available before the end 1993. The aim will be the adaptation to the new requirements and the reinforcement of the existing EC cooperation structures in developing countries with the view of making them more operational and action oriented. Furthermore, the establishment of new structures of an operational nature – Energy Centres in developing countries – are under consideration.

Such "Energy Centres" could cover a broad range of functions as, inter alia, the establishment of a clearing house mechanism for information dissemination, the promotion of demonstration projects, the preparation of cooperation programmes and assistance for financing investment projects. Such aspects should first be defined in a dialogue with a limited number of representative developing countries which could also be used as test cases for a more general strategy.

III.   Specific EC initiatives

A.   The Singapore Centre

The Singapore Authorities proposed the transfer of European environmental technologies to the region through the establishment of an EC-Singapore regional institute of environmental technology.

The Commission of the EC assigned a team of environmental experts to undertake a pre-feasibility mission to Singapore in May 1991. In 1993, a centre of excellence was set up. It is expected to become self-sustainable.

The main functions of the centre include:

–   technical services
–   human resources development
–   information services
–   policy development services.

The total cost to establish and operate the centre over an initial 3 year period is estimated at about 5 million ECU.

B. The Creation of a Euro-African centre for the transfer of environmentally sound technology

In a joint session on February 17, 1993, the French and German Ministers of the Environment proposed the creation of a Euro-African centre for the transfer of environmentally sound technology to African countries. It was also proposed that this initiative should be placed in the EC cooperation context.

The Commission of the EC is currently considering this proposal with a view to the possible participation of the European Community in its materialisation.

5. FINANCIAL RESOURCES

The Community together with its Member States have long been the largest source of official development assistance. In 1991 the Community plus Member States provided almost 50 per cent of all development assistance of DAC/OECD countries, equivalent to 0.42 per cent of the 12 Member States' combined GNP. In 1991, the Community's plus Member States' total ODA reached the level of 20.138 million ECU [1] of which 15 per cent was purely Community aid managed by the Commission.

Community aid activities

In 1992, the Community aid alone reached the level of 4 billion ECU. or the equivalent of 5.2 billion US $.

1) The situation in 1992

From a preliminary analysis of projects and programmes agreed during 1992, approximately 115 million ECU was committed for projects with primarily environmental objectives in ACP (Africa, Caribbean and Pacific) countries. The most significant interventions were in the following areas:

- water management
- environmental education
- integrated rural development
- biodiversity
- forest conservation.

The first projections for 1993 suggest a significant increase in "environment" projects and programmes,

[1] 1 ECU = 1.29 US dollars in 1992

particularly in relation to forest management and linkages with the health sector.

As regards projects and programmes in Asia and Latin America, approximately 10 per cent of technical and financial assistance resources is to be devoted over 1992-96 to projects and programmes with a primarily environmental objective. A first analysis of those actions approved in 1992 suggests that approximately 110 million ECU was committed during 1992 for direct environmental projects or projects where environment was among their objectives. The key areas of intervention were:

- provision of water
- rehabilitation of arid lands
- integrated rural development
- institution-building
- forest management and conservation.

In the Mediterranean, 115 to 120 million ECU have been foreseen for funding of environment projects and programmes in the period 1992-96, either for direct actions (pilot projects and training) or for interest subsidies for the loans of the European Investment Bank provided to Mediterranean Third countries in the area of environment. In addition to this amount, these countries benefit from the bilateral financial protocols which also include, in most cases, a considerable environmental component.

As with the Lome countries, the first projections suggest a significant increase in funding for "environment" projects and programmes in 1993.

2) Financing for Agenda 21 (1993)

A considerable part of the financial resources *available in* 1993 is expected to be committed for financing programmes and projects in various areas of Agenda 21. In particular, a total amount of about 770 million ECUs is expected, following initial assessments, to be committed in the following areas:

Table I (in million ECU)

Energy	69.0
Biodiversity	36.0
Forests	96.7
Desertification	11.0
Water	154.0
Rural Environment	270.0
Urban Environment	30.0
Human Development	68.6
Other	35.0
	770.3

148

(This table relates to Community financing under the Lomé Convention, the programme with the developing countries of Asia and Latin America, the Mediterranean aid programmes and other items in the Community budget)

It should be underlined that there are interlinkages between those areas: for example, numerous forestry, water and rural environment-related actions include a strong anti-desertification component. Forests will also benefit directly or indirectly from some actions classified under "rural development".

The total figure represents not only the financial resources to be committed in 1993 within the framework of development aid activities, but also incorporates the external component of other relevant Community programmes such as in the area of research, environment and energy, taking into account the part of the relevant programmes that is allocated to cooperation with developing countries.

While the bulk of the Community's environmental expenditure in developing countries is channelled through the major bilateral programmes, there are also two budget lines specifically devoted to environment and development issues. One, for the environment in developing countries, received a threefold increase to 26 million ECU in 1993. It is being used for pump-priming projects, for methodological research and to improve the Commission's capacity to integrate the environment into its own activities and in particular to reinforce Environmental Impact Assessment procedures. It is also being used extensively for formulating policy. The most significant proportion of the budget line will be devoted to institution building including training in environment management and establishment of sustainable development plans and strategies.

The second specific budget line concerns the protection of Tropical Forests. This line, which was initiated in 1991 with a budget of 2 million ECU, amounts to 50 million ECU in 1993. It is used specifically for the protection and rational management of tropical forests and the improvement of the conditions of forest dwellers and of populations living on the fringes of the forest.

NGOs play an important role in the execution of programmes and projects financed by these two budget lines.

Part of the above-mentioned resources will be committed to specific projects and programming in key Agenda 21 sectors for the early implementation of Agenda 21 as the Community's contribution to the 3 billion ECU initiative undertaken by the Community and its Member States at Rio.

CONCLUSIONS REACHED BY THE DEVELOPMENT COUNCIL ON 25 MAY 1993

In view of the first substantial meeting of the Commission on Sustainable Development (CSD) to be held in New York from 14-25 June 1993, the Council discussed several aspects of the follow-up to UNCED.

Following the global financial commitment made in Rio, the Council confirmed that the Community and its Member States would provide an initial tranche of 600 MECU in 1993 for specific projects and programmes in Key Agenda 21 sectors. In addition, the Community and its Member States will on a best efforts basis provide an extra 20 per cent (of 600 MECU) in new and additional resources.

The Council also took note that expert meetings had taken place between the Commission and Member States in order to identify initiatives and possible programmes, projects and joint actions and to consider areas for cooperation between Member States and the Commission. The Council welcomed work achieved so far and agreed that this shall be continued. Furthermore, the Council took note of information transmitted by the Commission which will form the basis for a factual report to the CSD.

CHAPTER XIII

ECONOMIC AND SOCIO-PSYCHOLOGICAL LIMITS OF SUSTAINABLE DEVELOPMENT: SOME "ENERGY SYSTEM" EXAMPLES*

Outline

I.   Introduction: the myth of a "sustainable development" paradigm

II.   **Actual developments:**

   2.1   exhaustible energy resources (physical limits)

   2.2   renewable energy resources (biological and life-cycle limits)

   2.3   the *economics* of physical and biological limits and the role of the Second Law of Thermodynamics in defining limits for closed systems (entropic limits)

   2.4   open societies: socio-psychological limits as generated by human value systems

   2.5   decision models and social structures

III. **Summary and extensions:** towards open systems and renewed growth: decision models for continued scientific search and renewal in an ever expanding universe...

---

*   Prepared by an ESCAP staff mumber as a discussion paper.

I. Introduction

At various times in human history, optimistic fads exited people: in the middle ages it was the *"philosopher's stone"*, turning everything into gold; or an *"elixir of life"* giving eternal youth.... In the "age of reason", it was the *"perpetuum mobile"*, a machine that would run forever..... In the late twentieth century, such a fad seems to be *"sustainable development"* that would keep societies forever young, increasingly prosperous, and always optimistic about the future.....

In the current fad likely to survive the turn of the millennium?

We argue below that we know enough already from both the physical and social sciences to answer the above question in the *negative*.

Does this condemn us to chronic pessimism? Not necessarily so.... we also know enough about the *limits* of our *knowledge* to be able to postulate at least two possible ways to remain optimistic:

1) a "sustainable stagnation", or culturally appropriate "survival" forms in the *closed* system of "spaceship earth";

2) an *open* system of outer-space, providing expanding spatial and time horizons for humanity (in an unrepent anthropocentric World......).

II. Actual developments

2.1 Exhaustible Energy Resources

Energy systems and development of national economies are intimately related, as reviewed before [1, 2, 4, 6]. Here we briefly quote from [4]:

"Development turns environmental resources into financial resources: that is how the poor (and the "not-so-poor") get rich! Where does energy come in? Stocks are turned into flows, then flows are turned back into stocks (money!) again and again. There are transaction costs "dissipating" accumulation during the process, producing waste and by-products. All these *transformations* require energy (activity, effort) – while the process itself seems to impose some sort of an *order* on the one hand (stocks of *final products*) while creating "confusion, waste, and distress" on the other (the garbage, the refuse, the waste, the slums, the unwanted by-products of a "used up" or *"converted"* environment!) at the same time. Is this process sustainable (*self-sustaining* would be a better word, or

"durable" as the French say it.....)? Should it be? As long as there is energy (activity, effort) there is no reason for the process to stop: there will always be sufficient "greed" in some members of society to continue turning environmental resources into financial resources, these financial resources will then *"sustain"* the process in the sense of "capital" for a *going concern.* Thus *"development"* will *not* go *out of business!* But are the final products, the order (and disorder) created really *desirable* by all constituencies affected by the process? Not *necessarily* so .... hence the current "environmental crisis" and the lipservice paid to "sustainable development" (a contradiction in terms!) that tries to achieve a delicate balance between the "goods" (getting rich) and the "bads" (getting filthy, sick and impoverished) of the process.

In figure 1 we depict the "triangular" relationship between Energy, Environment and Development."

In order to understand the meaning of this figure, let us quote again, now from [1]:

"Why are we reviewing energy resources? We have previously established that large flows of appropriate forms of energy would be required to meet development and welfare related needs of the South-East Asian countries. It is natural to ask, then: are stocks of energy commodities available to sustain the required energy flows? Is the mobilization of these energy flows sustainable, both in terms of the known and expected level of stocks of energy commodities, and, also, in terms of the environmental degradation (and increase of entropy) resulting from the conversion of "stocks" into "flows" at the substantial levels envisaged.

Energy flows (similarly to income flows) are also an indicator of "welfare". Welfare is a "flow" rather than a "stock" concept. For the energy or welfare situation not to worsen, by definition, it should, at least, be sustainable.

Energy resources in terms of the most common energy commodities (coal, oil, gas, biomass, uranium, etc.) could be considered similar to "wealth", (a stock concept!) not necessarily connected with welfare, unless converted into "flows"! Inactive capital (in terms of for instance hoards of gold by a miser!) is just as useless for welfare as mountains of coal (unmined!) or large accumulations of natural gas (unutilized!).

Technological processes (utilizing capital funds and human skills) would turn stocks into flows. For wealth, an elaborate system of circulation through the business and governmental redistribution activities

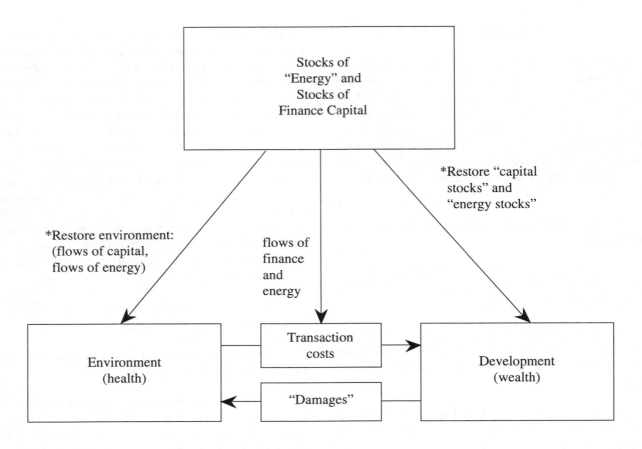

Figure 1

* These are reverse flows that have to be designed and implemented by society: *not* automatic!

would turn "wealth" into "welfare", that is into income flows for the participants of the economy, utilizing the "technology" of banking and finance capital, among others. For energy, the various supply choices (oil, gas, coal, nuclear, hydro, biomass) would create the energy flows, utilizing conversion technologies.

Note that just as in the case of wealth, sustainability for energy systems would mean that eventually "stocks" would have to be restored! This points toward the need for renewable resources, reforestation, recycling, etc. – even if it is hard to imagine for stocks of "wealth" such as oil reserves, coal reserves, uranium reserves! In case of hydropower and solar energy we only convert "flows" to "flows" (e.g. electricity), sustainability there means replacement of the conversion technology (dams, PV cells, etc.), a case similar to a certain extent to nuclear energy – although there the additional complication arises of securing stocks of fuel and disposing, stocks of waste!

What is clear, however, from the above discussion is that:

a)   resources (a "wealth" concept) cannot be considered in isolation of conversion technologies;

b)   sustainability of increasing energy flows depends on resources (in a complicated and indirect way) through supply chains that involve not only physical energy flows (and their consequences) but, also, the sustainability of economic systems that restore capital stocks (for conversion technology)."

Looking now again at Fig.1 in the light of the above, it becomes obvious why does the "sustainability" concept look dubious in the presence of exhaustible resources: income concepts (flows) and wealth concept (stocks) are linked in a fashion that does not give rise to automatic adjustments! Managed flows are required, welfare depending on goal-setting in all the stocks (capital, environment) resulting in forward and backward *flows ("incomes")* that are, in fact the measure of welfare of the participants in the process.

Contrary to the usual "automatic" adjustments of everyday economics, welfare (including environmental welfare!) is always a management dependent concept, empirically achievable (or not?) in any given society and time-horizon based on both the endowments and the skills existing in the participants in organizing the "flows" required to render continued (or may be even increasing?) welfare possible....

2.2 Renewable energy resources and "life-cycle limits" of economics

Although we concentrated above on exhaustible resources, the discussion can be easily extended to renewable resources through the *life-cycle* theory of conversion equipment (dams, solar cells, fast growing tree plantations, etc.). To analyze such conversion processes, we quote not quite extensively from [2], describing how the *valuation* process in economics is related to development and resource conversion:

"The science of economics has always searched for equilibria: supply and demand determine a market price that leads to a "balance". Such a price has acquired almost mythical qualities: in the middle-ages it was called the "just" price: if someone bought *below* it, he "won", if above it he "lost"... Marx has also asserted that if such "surplus value" is appropriated by the capitalist then in the wage-bargain the capitalist wins, while the worker looses: he is being exploited. Marx has conveniently overlooked the fact that if the worker values "leisure" higher than his wage, he would work less, in effect "exploiting" the capitalist through under-performance! The supply/demand of labour roughly determines wages, however who gets a better bargain is always an open question and depends, among other things, on individual evaluations of the deal by the various participants. Thus any exchange *must* lead to a *"surplus value"* for both the buyer and the seller in terms of their *own valuation* of the deal, otherwise the *exchange would never take place!* "Prices" are just an "accident", where we happen to agree so that *each* side might feel simultaneously that they had "won"!

The above analysis can be applied not only to simple exchanges as described above. It could also be applied to the "exchange" of converting capital into revenue, requiring only slight modifications. What are we talking about?

Take, for instance, a ship as an example (we are talking about islands, after all...). The ship costs, shall we say, one hundred thousand dollars. The owner acquired it so that he would have some revenue from it: let us assume he would like to have 10,000 dollars every

year during the next ten years. He may thus lease out the ship for 10 years to someone who would pay him 10,000 dollars per year for the use of the ship. The one who pays this user's fee is concerned only that among all his expenses this 10,000 dollars would not represent an inordinately large portion: one must eat, one must drink, one must live – not only carry goods on the sea.... (We do not ask at present where the original capital is coming from – we will return to that later.) What we wanted to illustrate here is that in the above exchange while one side's interest is a strict "revenue" objective, the other's interest is acquiring a service within a "budget constraint".

If the lessee acquires the ship below 10,000 dollars, then he wins: he can use his budget for other purposes after having transported all the goods he wants transported. If the lessor gets more than 10,000 dollars, he wins: his capital will give him more revenue than what he expected.

For how many years is the ship "good"? If it lasts exactly 10 years, then both the lessee and the lessor achieve their expectations. If it lasts longer potentially both can win...How? It is obvious for the lessor, however the lessee may gain also: the lessor may give him a discount for these late years, or even keeping the rent constant might represent a net gain to the lessee in an inflationary environment. If the ship sinks before the 10 years are over they both may loose – especially the owner, but the lessee, too, by having to "shop around" again for already contracted services of a ship.

When they first reach an agreement neither of them know this: the future is always uncertain. For instance, if they knew that the ship is "good" for 20 years the lessor may let it at 5,000 dollars per annum on a 20 years contract, giving all the "technical advantage" to the lessee, or for *anything* between 5 and 10,000 dollars, dividing up the benefits of technical advantage according to the agreement reached.

The above train of thought has *not* taken into account the "time-value of money". Implicitly we have assumed that the 10 year payback period compensates the owner of capital. If he gets a higher rent, his payback is shorter, if he gets a lower rent, his payback is longer. The lessee would always want longer time periods and lower rates. Agreement depends on whether they assess correctly the "lifetime" of the asset and their own valuation of a dollar *today* in their pocket (as opposed to only *next* year) is worth how much more? If we analyze the time-preference of each we may find that such totaled up rates of return at their own valuation (discount rate or time value of money) might result in a

156

similar asset price (future value of asset) at a particular lease payment – and we get a graph something like that depicted in Fig.2. Here, say at 12,000 dollars annual lease both lessee and lessor take the value of the ship to be 200,000 dollars. From the original asset price (and implicitly assumed zero time value of money) if follows that at 100,000 dollars (as long as the lessor assesses the lifetime of the asset at 20 years!) there is a wide "gap" where both lessor and lessee would be satisfied while at 200,000 asset price ("future value", taking the time preference of each into account) the possible agreement shrinks to a point! Economists would rejoice: we found an equilibrium, a "fair" valuation of assets satisfying everybody.... Is such an equilibrium in fact desirable?

As seen on the Figure, while in the shaded area several agreements are possible where both lessor and lessee are satisfied, on the right hand side of the crossing of the lines only one *or* the other can be satisfied as long as they stay on their line – but never both! Thus, on the right-hand side, one would argue that assets are over-valued, one can talk of "exploitation" of either the lessee by the lessor (upper line) *or* the lessor by the lessee (lower line).

Let us now assume that we are no longer talking about a ship, but, let us say, about an oil-field or a coal-mine. Then the asset value will compensate the owners during the exploitation process (over the lifetime of the mine or field): the capital returns in the form of sales revenues on oil or coal so that it can be used again for the exploitation of new coalmines or oilfields.

What happens if assets are overvalued, that is if prices rise too high, the seller of the natural resource wants more than the exploitation costs of the oilfield/ coalmine (including the "normal rates of return" of the industry on the one hand, and the consumer's "time value of money" on the other, the two lines, as depicted in Fig 1)? It is likely that we end up to the right hand side of the intersection of the lines, no-one would be satisfied, they will be angry at each other, shouting "exploitation" and the economy would begin a downward "adjustment" of asset values (or recession, depression, etc. in everyday terms!).

While asset prices are such that we are at the left hand side of the intersection point, they both get more than what they expect, there is a general climate of

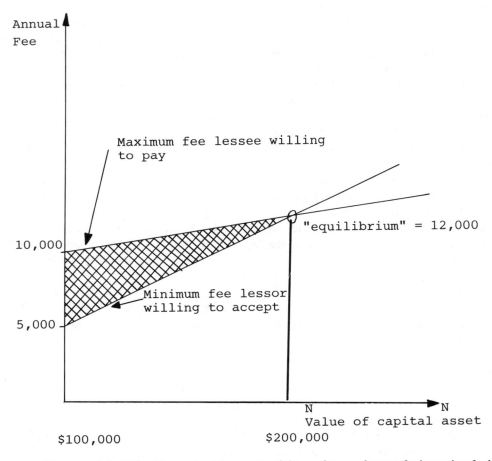

Note: This graph assumes 20 years technical life of the asset and lower rates of time preference of money for lesses than for lesser, "totalled up" as "future value" to the "equilibrium" that is the annuity equivalent of the income stream.

Figure 2.

optimism, economies develop further and further (a "boom" in everyday terms!) in a spirit of mutual goodwill....

As we see, to the chagrin of the economists, "equilibrium" here is the point where "boom" turns into "bust", "just price", "justice" is a most *unstable* equilibrium point....

Going further: what would happen now if we slam an environmental tax on oil/coal prices? Again, it is likely that we push asset values to the *right* of the equilibrium point, we get a "vicious society" where everyone blames everyone else end start a *downward* spiral.

This is the economics of environmental protection – unfortunately very few people of goodwill and genuine environmental concerns (conservation of resource base – a laudable objective!) are yet aware of the repercussions of "absorbing environmental costs"! In a *closed* system, such a course could lead to disaster....

2.3 The role of the second law of thermodynamics

To go even further and establish entropic limits for closed systems, we quote further from [2]:

"Continuing the above train of thought a little further: how is capital formed (i.e. the capital that the owner of the ship has invested into his vessel)? Originally in most cases some natural resource (oil, coal, etc.) has been turned into financial capital by the process described above. The whole banking system, discounted present values, future values, lending at interest, investing at "normal rates of return" has been devised to safeguard financial capital from dissipation once it has been formed. Capitalization thus increases with development and thus, by necessity, the "weight" of original natural resources will be less and less in more developed national economies in value terms. However, there is another basic consideration concerning "open" and "closed", production systems that has fundamental importance not only for island economies but also for the world [Fig. 3].

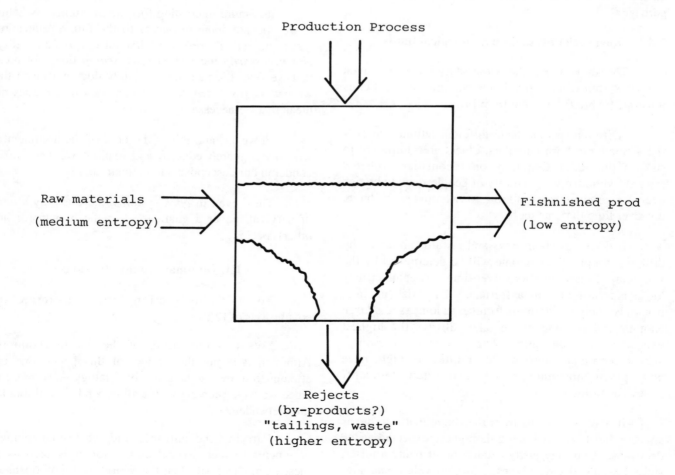

Figure 3

As depicted on the figure, all production processes (mines, factories, etc.) work over some raw materials producing some "finished product". From physics, an intervention of this sort causes a "medium entropy" (slightly disorganized) raw material to turn into highly organized (lower entropy) finished products. In the process, however, there are rejects, bi-products, waste – a higher entropy "garbage" stream toward the bottom. The second law of thermodynamics says that after each intervention entropy increases, thus, by necessity (since we know finished products to be low entropy) the "garbage" stream is higher entropy than the original raw material. In a closed system, only this "garbage" is left as raw material for future production processes, thus production in closed systems gets us closer and closer to "entropic death" – irrespective of whether the final products can be sold or not! (as described in the previous section in terms of "values" and "exchanges"). Higher asset values are thus a necessary consequence also of environmental *degradation* in closed systems....more complicated, longer, more expensive production processes working on raw materials that consists of more and more recycled garbage!"

2.4 Open societies: socio-psychological limits

The above is *not* the whole story however. Even in open systems, socio-psychological limits exist. Let us illustrate by briefly quoting from [3]:

"Different people have different attitude towards risk. Some are born gamblers, others are indifferent to risk in the sense that they only consider *expected* rewards while the vast majority of people seem to be risk averse: we want to hold on to whatever we have, however little that may be!

If decisions about risks and rewards are taken by different people, the outcome will be determined by the sum-total of these decisions over time. A gambler may loose his shirt or win a fortune. The "risk neutral" person is likely to retain his fortune, as long as it is *large* enough and as long as he takes always the highest expected value decision. The risk-averse person is almost certain to loose whatever little he might have now, given long enough time and a few "prudent" miscalculations.

It may be obvious from the above that I argue, as an economist, for risk-neutral, highest expected value decisions. You may justly excuse me of being a miser, since I would decide highest expected value outcomes preferable no matter how large my fortune already is.... on the other hand, if I am a gambler and am faced with

ruin, standard references are available from the literature on probability calculus that "to risk all you have" (or "almost") may be the optimal strategy, assuring highest *probability* of survival (however small that may be!). For risk-averse societies, eventual extinction is an *almost* certainty given long enough time and a series of sub-optimal decisions."

2.5 Decision models and social structures

I have analyzed these questions before in terms of some decision models (5, 6).

In Fig. 4 we summarize such a decision model.

What is seen here is that at each point in time resources are allocated to long-term "gambling on the future" (research, exploration, search processes) and immediate gratification (wages, consumption). It is obvious that in such a two sector model, if nothing is allocated to search processes, the "sustainability" of the system is limited to one time period: after we consume all we have, we die....

It is more interesting for "sustainability", whether if we allocate *some* resources to the future (education, research, exploration) will that ensure sustainability? Not *necessarily* the fact that we *search* does not mean that we find, the fact that we study does not mean that we necessarily *learn,* the fact that we *explore* does not mean that we necessarily *discover*!

However, and this is the crux of the argument, if we do *not* search educate and explore we *necessarily* condemn our descendants to extinction....

Thus, even in *open* systems, there is only a hope of survival, never a guarantee, for the human (or any other) species....

III. Summary and Extensions

To start this section let us re-iterate the conclusions of [2]:

"What is the future of the World economy? Although it is not the subject of this paper, one can attempt to draw a conclusion by analogy – the more so since we have seen above we all depend on it through " interdependence".

Carrying the entropic and waste-management arguments to their logical conclusion it is seen as an inescapable necessity that the World economy opens up "entropic sinks" in outer space – otherwise we all suffer "death by constipation" of closed systems.

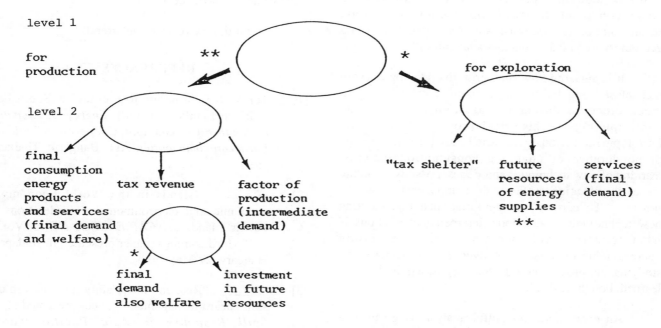

```
level 1

for                    **              *              for exploration
production
```

level 2

final
consumption
energy   tax revenue
products   factor of   "tax shelter"   future   services
and services   production   resources   (final
(final demand   (intermediate   of energy   demand)
and welfare)   demand)   supplies
                                                                             **
                        *
                       final   investment
                       demand   in future
                       also welfare   resources

           *   feedback loop of future demand for exploration investments

          **   feedback loop of future total production demand based on success of previous exploration

Figure 4. Current energy investments – determinants of demand levels

This is admittedly speculative. Still we wish to close on a hopeful note. Environmental problems will be solved by developed countries through space research and "entropic sinks" found and developed in outer space, allowing, thus, further development of all the "islands" of the developing world...."

Finally to put all this in a socio-psychological and historical context, we close with our earlier analysis in [6]:

"In a Lockian political system (small), quasi-independent politico-economic units trading with each other, e.g. small farmers and merchants) an Adam Smith type competitive market system is deemed to lead to "highest efficiency" in the sense of most wealth produced by specialization according to comparative advantage and free and unrestricted trade among the politico-economic units, increasing thus possible consumption choices. The system does not produce equality, but rewards individual effort.

An ideology results which tries to reinforce such a self regulating system: we have conventional market theory and partial equilibrium analysis of economic problems.

Consumption needs are taken for granted, and the production system develops so as to best meet consumption needs. There is no higher purpose than maximum efficiency in satisfying consumption needs and/or producing maximum yield on invested capital. A "general equilibrium" is reached which achieves maximum welfare consumption.

Changing tastes and changing technology are exogenous: markets would adjust so as to accommodate these.

In a Hobbsian political system, on the other hand, participants of the nation-state (hierarchically structured in terms of their role as "governed" and "those who govern") are assumed to need guidance by the state in knowing what is "good for them". Markets are assumed to cater for "perverse needs", and to facilitate "exploitation" by the propertied classes (of those who are their "vassals"). A Marxian economic system would thus concentrate on regulation, to achieve a political economy "where each contributes according to his abilities and each and every member can satisfy his/her needs". The difficulty with the latter system is that in addition to assuming well-defined needs and abilities it also assumes omniscience by the regulators of the politico economic processes in order to achieve satisfaction for all. In practice this doesn't always work, so the solution is: you must want (and love!) what you get, otherwise you are an anti-social outcast!

It is amazing how such primitive ideologies (with their closed, self-reinforcing systems) have achieved so much credence during historical times. Anybody contemplating the historical process of Economic Development seriously is struck, first of all, by the preponderance of changing consumption patterns, learning of new techniques and trades, possibly changing social stratification as the main components of such a process. "Unlearning" in-appropriate patterns, adopting new technologies, behaviour, interpersonal relations is what renders development possible. A certain understanding of what is involved can be achieved by studying the process how technology is acquired, trade, learned, bought and sold.

An even more thoughtful analysis is given in a study related to the "Absorptive Capacity" concept. The point is raised in a latter study that conventional concepts of "absorptive capacity" limiting economic development need to be revised: economic development itself must be defined as involving possible changes in absorptive capacity."

As a final note: the socio-psychological limits of development in modern societies may work, through public opinion eventually influencing the electoral process toward a "lowest common denominator", risk-averse group of leaders who, in turn, will then dissipate society's wealth ("the wealth of nations") through redistributive schemes that emphasize present day "fire fighting" rather than investment in search processes for future options. Ambitions space programmes, for instance, are thus scrapped, inadvertently starting thereby a downward spiral of expectations, that leads to a shrinking of the available credit volume and a loss of business confidence. That, in turn, puts into deficit "social safety net" programmes, just at the very time these would be most needed....

Isn't it time to reverse such trends?

REFERENCES

[1] Török, S.J. "Energy Resources in South-East Asia" in Lapillon, B. (ed): *Energy Development in South-East Asia and Cooperation with the European communities,* AIT, Bangkok, Thailand, 1990.

[2] ———— , "Islands in the World Economy – economic and environmental considerations in Pacific Islands", *ESCAP ENERGY NEWS,* Vol.X, No.2 (forthcoming) (paper presented in Budapest, Hungary, 27 May 1993).

[3] ———— , "Risk and uncertainty trade-offs in oil-spill hazard management" (paper presented at *Oil Spill Response in Asia Pacific Waters* Conference, Kuala Lumpur, Malaysia, 30 June - 3 July 1993).

[4] ———— , *"Simple Environmental Economics of Asian Coal and Gas",* paper presented at Asian Development Bank, Manila, Philippines, March 1992.

[5] ———— , *Value Changes, Gaming, a Behavioural Approach to Guided Negotiations and Macroeconomic Planning,* Ph.D. Dissertation, Columbia University, N.Y., 1976 (Xerox Microfilm Systems, Ann Arbor, Michigan, 1978).

[6] ———— , "Fueling Growth: the relevance of the Japanese model to Eastern Europe, in Izawa, Y. (ed.): *Asia-Pacific Symbiosis – Quest for Business-Economic Reciprocity,* Sohsei-sha, Tokyo, 1992.

[7] ———— , Journal of the Siam Society, 1993.

CHAPTER XIV

GLOBAL WARMING, RELATED CO_2 REDUCTION PROPOSALS AND CONSEQUENT IMPACT ON ENERGY POLICY*

1. Introduction

Nothing in human history have ever been as dose as a truly common problem for mankind as global warming. Since scientists' Villach meeting on global warming and its consequent climate change in 1985, tremendous international efforts have been done to study global warming and to establish the action plan against it in an unprecedently short period.

The UN Conference on Environment and Development is scheduled for June 1992 in Rio de Janeiro, Brazil, where a framework convention on green house gases reduction is supposed to be adopted. Judging from the current inertia of worldwide passion on global warming issue, it is quite possible that several protocols will follow the convention immediately.

There is no doubt that developed countries are responsible for current global warming phenomenon due to their massive accumulated CO_2 emission since the Industrial Revolution. At present developed countries account for 2/3 of total CO_2 emission. But the situation might be reversed in the not so distant future. Developing countries might contribute at least 60 per cent of world CO_2 emission within 20 years due to the multiplying effect of rapid population and economic growth. Without full participation of developing countries any effort to stabilize atmospheric CO_2 concentration would be destined to fail.

The most important single greenhouse gas (GHG) is carbon dioxide. Thus, control on CO_2 emission is inevitable. This necessarily brings about a serious constraint on fossil fuel consumption and consequently on whole energy sector. Energy is so essential to economic growth to achieve better standard of living that any developing country cannot accept a compulsory measure to out down fossil fuel consumption without having reasonable alternatives. Some still argue that global warming is just a plausible scientific hypothesis. Obviously there is a strong potential conflict between the

North and the South. Therefore it is critical to have a common view at least on the relationship between atmospheric CO_2 concentration and fossil fuel comsumption before we discuss how to allocate the responsibility to cut down fossil fuel derived CO_2 emission.

So far it has been quite often mentioned that CO_2 emission should be allocated based on various criteria such as population, current emission level, GDP, or a formula incorporating these factors. But it has been seldom told what each allocation scheme really means.

Therefore, the purpose of this paper is to draw a common observation on the relationship between man-made CO_2 emission and atmospheric CO_2 concentration, to analyze the factors that affect CO_2 reduction allotment and the impacts on future energy demand and supply, in order to assist energy policy makers both in developed countries and in developing countries understand the fundamental constraint on energy sector resulting from global warming related CO_2 reduction and hopefully have a common objective starting point to deal with global warming negotiation in energy sector and to investigate the optimum stabilization goal and process acceptable to all soverign countries, based on equity and applicability.

II. Man-made CO_2 Emission and Atmospheric CO_2 Concentration

Man-made CO_2 mainly comes from fossil fuel burning. By-products from cement manufacturing consists relatively small portion, less than 3 per cent of total anthropogenic CO_2 since 1959. Table 2.1 shows man-made CO_2 emission from 1959 to 1988.

The pre-Industrial Revolution atmospheric CO_2 concentration was estimated at about 278 ppm from the Antarctica ice core record (US DOE, 1990). Direct measurement of Atmospheric CO_2 concentration began in 1958 by the U.S. Mauna Loa Observatory. Figure 1 shows atmospheric CO_2 concentration since 1959 (UNEP, 1989). It is observed that atmospheric CO_2 concentration has increased and the increasing rate is being accelerated since the first direct measurement.

* Prepared by Jung-Sik Koh, Regional Adviser on Energy (1991-1993), ESCAP for a meeting on research on greenhouse gases organized in August 1991.

Table 2.1 Global carbon dioxide emission

(unit: 106 Ton C/yr)

Year	Soild Fuel	Liquid	Natural Gas	Gas Flaring	Cement Production	Total
1958	1 344	732	192	35	36	2 339
1959	1 390	790	214	36	40	2 470
1960	1 419	850	235	39	43	2 586
1961	1 356	905	254	41	45	2 601
1962	1 358	981	277	44	49	2 709
1963	1 404	1 053	300	47	51	2 855
1964	1 442	1 138	328	51	57	3 016
1965	1 468	1 221	351	55	59	3 154
1966	1 485	1 325	380	60	63	3 313
1967	1 455	1 424	410	66	65	3 420
1968	1 456	1 552	445	73	70	3 596
1969	1 494	1 674	487	80	74	3 809
1970	1 571	1 838	515	88	78	4 090
1971	1 571	1 964	553	90	84	4 244
1972	1 587	2 056	582	95	89	4 409
1973	1 594	2 240	608	112	95	4 649
1974	1 591	2 244	616	107	96	4 654
1975	1 686	2 131	620	96	95	4 628
1976	1 723	2 313	644	110	103	4 893
1977	1 786	2 390	645	108	108	5 037
1978	1 802	2 384	672	106	116	5 080
1979	1 899	2 535	710	74	119	5 337
1980	1 921	2 409	721	78	120	5 249
1981	1 930	2 274	730	58	121	5 113
1982	1 986	2 188	724	56	121	5 075
1983	1 992	2 167	730	52	125	5 066
1984	2 080	2 200	783	47	128	5 238
1985	2 173	2 182	807	46	130	5 338
1986	2 250	2 297	827	45	136	5 555
1987	2 615	2 100	910	n.a.	165	5 790
1988	2 707	2 183	935	n.a.	173	5 998

(Source: 1958-1986, UNEP Environment Data Report, 1989
1987-1988, Estimated from UN Energy Statistics Yearbook, 1990)

The global carbon cycle is such a huge and complicated system that it cannot be quantitatively well defined. A lot of work has been done to describe the global scale CO_2 behavior. The main natural factors affecting the carbon cycle are mass transfer between atmospheric CO_2 and ocean surface water, uptake by photosynthesis and release by respiration.

Atmosphere is homogeneous upto 100 km. It is well mixed and does not show any composition variation. Table 2.2 shows atmospheric properties vs. height from the sealevel (Lapedes, 1977).

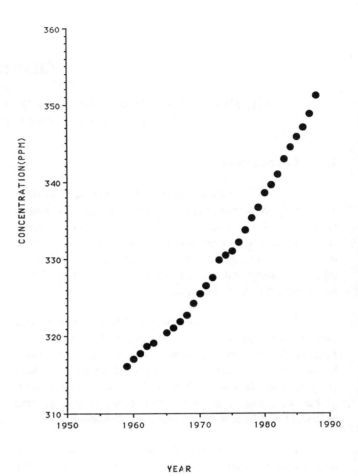

Figure 1. Atmospheric CO_2 concentration

Table 2.2 Atmospheric properties

Height (km)	P (mb)	T (°K)	Density (kg/m³)	M.W. (g/gmole)	Layer
0	1 013	288	1.23	28.96	Troposphere
5	540	256	0.736	28.96	
10	265	223	0.414	28.96	
20	55.3	217	0.0889	28.96	Stratosphere
40	0.287	250	0.004	28.96	
60	0.225	256	0.000306	28.96	Mesosphere
80	0.0104	181	0.0002	28.96	
100	0.0003	210	0.000000	28.88	

Meterial Balance for atmospheric CO_2 is given as follows:

$$\frac{M.W. (CO_2)}{M.W. (Air)} \times \int_{h=O}^{h=H} \rho\,(h) \times 4\pi\,(R+h)^2 dh \times \frac{dC}{dt} = i - 0 \quad (Eq.2.1)$$

where,   C :   Atmospheric CO_2 Conc.   [ppmv]

         H :   Upper atmospheric limit   [km]

         M.W. (CO_2) :   Molecular Wt of CO_2 [g/gmole]

         M.W. (Air) :   Molecular Wt of Air   [g/gmole]

         R :   Radius of Earth   [km]

         h :   Height from the earth surface   [km]

         i :   Input of CO_2 into Atmosphere   [bil.Ton C/yr]
such as fossil fuel combustion, cement manufacturing, land use change, volcanic eruption, etc.

         o :   Output of CO_2 from atmosphere   [bil.Ton C/yr]
such as absorption by ocean, uptake by terrestrial ecosystem

         t :   Time   [yr]

         ρ :   Density of Air   [kg/m³]

Assuming air density changes linearly between the interval in Table 2.2, integrating the pre $-\dfrac{dC}{dt}$ term of Equation 2.1 from h = O to h = 60 km, we get the following expression and the numerical estimate of M as 2.39 bil.Ton C/ppm.

$$M \times \frac{dC}{dt} = i - o \qquad (Eq. 2.2)$$

Where, M :   Carbon Equivalent Mass per Unit Change of Atmospheric Concentration   [bil.Ton C/ppm]

Not only the man-made emission but also the natural emission contribute to the CO_2 influx into the atmosphere. Furthermore, there exists a natural disturbance term like volcanic eruption. Output term is even worse than Input term to be described quantitatively.

Therefore, reconceptualizing the equation as follows:

$$M \times \frac{dC}{dt} = i^* - An \qquad (Eq. 2.3)$$

Where,   i* :   Man-made Emission

        An :   Apparent Natural Absorption

An is defined as the difference between man-made emission and increase in atmospheric CO_2.

To decrease the impact of natural disturbance term the above equation can be better expressed as an integrated form for the interval $t = t_0$ to $t = t_1$.

$$M \int_{t_o}^{t} \frac{dC}{dt} = \int_{t_o+1}^{t} i^*dt - \int_{t_o}^{t} Andt \qquad (Eq. 2.4)$$

Assuming natural absorption is constant for the integral time period, we get the following expression.

$$An = \frac{1}{t - t_o} \left\{ \int_{t_o+1}^{t} i^*dt - M\,[C(t)-C(t_o)] \right. \qquad (Eq. 2.5)$$

Applying the data from Table 2.1 and Figure 1 to the derived expression, apparent natural absorption An and remaining ratio R of man-made CO_2 emission were estimated since 1959.

Three observations can be obtained from Table 2.3. First, the rate of increase in atmospheric carbon concentration has been accelerated along with increased fossil fuel consumption since the atmospheric CO_2 concentration was directly measured in 1959. Second, approximately 60-70 per cent of man-made CO_2 emission has been apparently added to atmospheric CO_2. Third, in relation to the second observation, the apparent natural CO_2 absorption capacity can be estimated as 1 - 2 bil.Ton Carbon per year for the past three decades.

Table 2.3 Apparent natural absorption & remaining ratio

(unit: BTC, ppm)

	'59-'65	'65-'70	'70-'75	'75-'80	'80-'85	'85-'88	TOTAL
Acc.CO_2	16.921	18.228	22.584	25.553	25.83	17.344	126.46
dCO_2	4.31	5.1	5.5	7.39	7.25	5.35	34.9
M×dCO_2	10.3009	12.189	13.145	17.6621	17.3275	12.7865	83.411
An	1.1034	1.2078	1.8878	1.6451	1.7005	1.5192	1.484448
R	0.6088	0.6687	0.582	0.6912	0.6708	0.7372	0.659584

M: Carbon Equivalent Mass per PPM = 2.39BTC (Billion Ton Carbon)
ACC.CO_2: Accumulated Emission, dCO_2: Atmospheric Concentration Difference

From the above postulated CO_2 material balance, we may reasonably assume there would be a direct relationship between atmospheric CO_2 concentration and accumulated man-made CO_2 emission. Accumulated man-made CO_2 emission vs. atmospheric CO_2 concentration data since 1959 were fitted by various regression equation.

As shown in Table 2.4, exponential regression fits excellently the data since 1959. Following relationship was obtained.

Table 2.4. Regression results

Type	Equation	R-Square
Lin	315.8806 + 0.0027 X	0.9983
Exp	316.1523 × exp (8.1290E-6X)	0.9988
Log	255.5415 + 9.1429 X lnX	0.7951
Pwr	263.4011 × X^{0.0276}	0.8039

$$CO_2 \text{ (ppm)} = 316.15 * \exp(8.1290E\text{-}6*X) \qquad \text{(Eq. 2.6)}$$

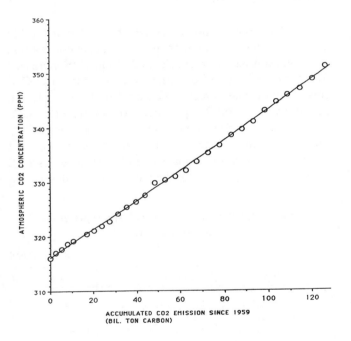

Figure 2. Accumulated CO_2 emission vs. atmospheric CO_2 concentration

III. SETTING THE GLOBAL GOAL AND ITS FAIR DISTRIBUTION

(1) Setting the Goal

If possible, CO_2 emission should be conceptually equal to or less than Natural Absorption Capacity.

Actually nobody knows how fast the reduction of CO_2 emission should be to prevent catastrophic consequences. What is certain is the fact that "The sooner, the better". Needless to say the Stabilization path should be politically acceptable to all the countries in this planet. Even though global warming were a solid scientific fact, the Stabilization path should be a technologically-economically feasible one without sacrificing economic growth especially in the developing countries.

Apparent Natural Absorption was defined at the preceding chapter as the difference between man-made CO_2 emission and atmospheric CO_2 increase. It is estimated as the range of 1.2 bil.Ton-Carbon and 1.9 bil.Ton-Carbon per annum. Therefore, idealistically current level of 6 bil.Ton-Carbon should be reduced to approximately 1/3 of present level. But, it would be a good idea to hold even this tentative goal until we finish reviewing the long term impact on future energy supply of the major proposed scheme to stabilize CO_2 emission in the next section.

(2) Review on the Major Proposed Scheme to Stabilize CO_2 Emission

Whatever the reduction level may be, it is most critical how to distribute the burden between the soverign countries.

Being the most fair single criterion, population prorated apportioning is not likely to be practicable. From the point view of developed countries, current emission or a GDP based allocation is more acceptable than the sole population based allocation. There is no doubt that a criterion would be much more favorable to a specific country than to an other.

Therefore, it is necessary to identify the factors affecting each individual country's interest explicitly. Only after we clearly understand the short-term and long-term implications on each individual country's interest, we could start the meaningful negotiation to reduce the CO_2 emission on a global scale.

If we define $X(t)_i$ as country i's Bau (Business as Usual) Carbon Emission in year t, $X(t)_i$ can be given as follows:

$$X(t)_i = P(t)_i \times GDP(t)_i \times E(t)_i \times C(t)_i \qquad \text{(Eq. 3.1)}$$

where, Population: $P(t)_i = (1+a) P(t_o)_i$ [mil.]
Per Capita GDP : $GDP(t)_i = (1+b) GDP(t_o)_i$ [U$]
Energy Use per GDP : $E(t)_i = (1+c) E(t_o)_i$ [TOE/U$ 1000]
Carbon Intensity : $C(t)_i = (1+d) C(t_o)_i$ [Ton Carbon/
in Energy Mix   TOE]

Then, global CO_2 emission Y(t) is the sum of emissions from all the countries.

$$Y(t) = \Sigma\ X(t)_i \qquad\qquad \text{(Eq. 3.2)}$$

Where, i : Country Code

t_o : Reference Year

t : Target Year

To give practical insight for the underlying difference between alternatives, all the countries are catagorized as three groups, Group A; Developed countries, including OECD, Eastern Europe and USSR, Group B1; NIES including Brazil, Mexico, the Republic of Korea, Taiwan Province of China, Singapore and Hong Kong, Group B2; other developing countries.

General framework of assumptions are summarized at Table 3.1.

Carbon Dioxide emission from energy sector was estimated until the end of 21st century based on the above assumption. Figure 3 and 4 show the result of proposed model simulation.

This implies that even though developed countries contribute about 60 per cent of global emission at the moment, developing countries would contribute majority of CO_2 emission in the near future.

But in terms of per capita emission, parity would be achieved only after the mid 21st century.

Table 3.1 General Framework of Assumptions

	1988 ~ 2000	2000 ~ 2010	2010 ~ 2030	2030 ~ 2050	2050 ~ 2070	2070 ~ 2090	2090 ~ 2100
P.GROWTH							
DEVELOPED	0.0047	0.0028	0.0028	0			
NIES	0.0166	0.0116	0.0116	0.0047	0.0028	0	
DEVELOPING	0.0203	0.0144	0.0144	0.0116	0.0047	0.0028	0
E.GROWTH							
DEVELOPED	0.027	0.027	0.01	0.01	0.005	0.005	0.005
NIES	0.05	0.04	0.0295	0.027	0.01	0.005	0.005
DEVELOPING	0.038	0.038	0.04	0.0295	0.027	0.01	0.01
	1988	2010	2030	2050	2070	2090	
E.EFF							
DEVELOPED	0.4181	0.27	0.2430	0.2187	0.1968	0.1771	
NIES	0.4663	0.4181	0.2700	0.2430	0.2187	0.1968	
DEVELOPING	0.6341	0.4663	0.4197	0.3777	0.3399	0.3059	
	1988	2020	2030	2050	2070	2080	2090
C.INT							
DEVELOPED	0.764	0.4879	0.4638	0.2843	0.2727	0.2669	0.2611
NIES	0.7447	0.6611	0.5953	0.4638	0.2843	0.2785	0.2727
DEVELOPING	0.8556	0.7965	0.7374	0.6611	0.5296	0.4638	0.3741

(1)   P.GROWTH : ESTIMATED from "UN WORLD POPULATION CHART 1990"

(2)   E. GROWTH
88-2005 Forecast : OECD 2.7 per cent, DEVELOPING 3.8 per cent (IEA, 1991)
OECD 80-88 Actual: 2.95 per cent, assumed for 2010-2020 NIES, 2030-2050 Developing countries (IEA, 1990)
OECD 60-80 Actual: 4.0 per cent, assumed for 2000-2010 NIES, 2010-2030 Developing Countries (IEA, 1990)

(3)   E.EFF : 0.27 ~ 1988 JAPAN Actual (BP, 1990, IMF, 1989)
Developed countries are assumed to achieve 1988's Japanese level by 2010.
NIELs are assumed to achieve same level 20 years later.
Developing countries are assumed to achieve 1988's NIEs level by 2010 and thereafter 10 per cent/20 year efficiency improvement is assumed.

(4)   C.INT : 0.4879 ~ 1988 FRANCE Actual (IEA, 1990)
Developed countries are assumed to achieve 1988's French level by 2020 through extensive nuclear option and to further decrease to 0.4638 by 2030 through gas penetration into transportation sector. Thereafter hydrogen fuel cell and solar energy are assumed to replace transportation and residential sector respectively. Twenty years technology transfer gap is assumed between developed countries and NIEs, again between NIES and other developing countries.

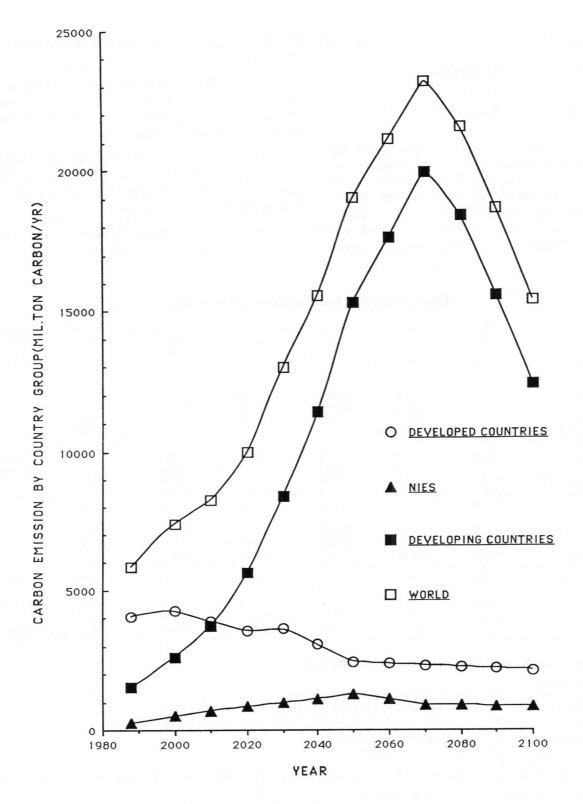

Figure 3. Carbon emission by country group

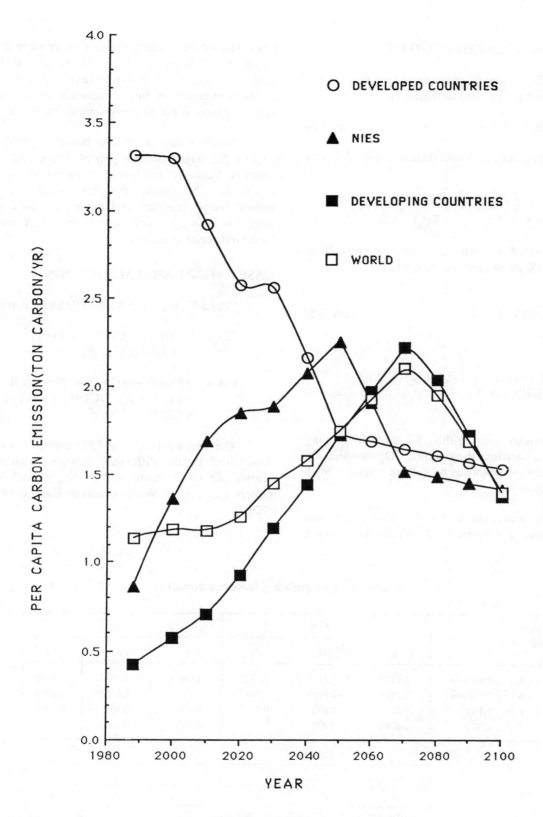

Figure 4. Per capita carbon emission

CASE I. CURRENT CARBON OUTPUT

If we define country's amount of sacrifice at year t as DX, DX can be expressed as (Eq. 3.3).

$$DX = X(t)_i - X(t_0)_i \qquad \text{(Eq. 3.3)}$$

Substituting (Eq. 3.1) into (Eq. 3.3), (Eq. 3.4) can be obtained.

$$DX = [(1+a)(1+b)(1+c)(1+d)-1] \\ P(t_0)_i \times GDP(t_0)_i \times E(t_0)_i \times C(t_0)_i \quad \text{(Eq. 3.4)}$$

Accumulated sacrifice can be obtained from integration of DX for the interval of t_0 and t.

$$X = \int_{t_0}^{t} DX \, dt \qquad \text{(Eq. 3.5)}$$

$$X = \int_{t_0}^{t} \{[(1+a)(1+b)(1+c)(1+d)-1] P(t_0)_i \times \\ GDP(t_0)_i \times E(t_0)_i \times C(t_0)_i \} \, dt \quad \text{(Eq. 3.6)}$$

It is obvious from Table 3.2 that developing countries would sacrifice their future CO_2 emission far more than developed countries by current emission level based stabilization.

It is well attributed to the fact that developing countries' future economic and population growth potentials would be much higher than those of developed countries'. Therefore, if we decide to stabilize CO_2 emission based on current level, it follows that developing countries' future economic growth would be seriously limited due to energy supply shortage.

Another aspect should be mentioned here to explain the huge potential gap of future CO_2 emission sacrifice between developing countries and developed countries. Developed countries would be able to reduce energy intensity and carbon intensity far more than developing countries due to their high technology based industrial structure.

CASE II. GDP BASED ALLOCATION

Similarly, Eq.3.7, 3.8 and 3.9 can be derived.

$$X(t_0)i = Y(t_0)/[\Sigma GDP(t_0)_i \times P(t_0)_i] \\ \times GDP(t_0)_i \times P(t_0)_i \qquad \text{(Eq. 3.7)}$$

$$DX = [(1+a)(1+b)(1+c)(1+d) \times E(t_0)i \\ \times C(t_0)_i - Y(t_0)/\Sigma GDP(t_0)_i \times P(t_0)_i] \\ \times GDP(t_0)_i \times P(t_0)_i \qquad \text{(Eq. 3.8)}$$

The amount of future CO_2 emission sacrifice is determined by the difference between current carbon output per GDP multiplied by the growth potential factors and current World Average Carbon Output per GDP.

Table 3.2. Determining factors estimation

Factor	2000			2010		
	A	B1	B2	A	B1	B2
1 + a : P.Growth	1.0579	1.2184	1.2727	1.0879	1.3674	1.4683
1 + b : E.Growth	1.3767	1.7959	1.5645	1.797	2.6583	2.2717
1 + c : E.INT	0.823	0.895	0.8677	0.646	0.79	0.7354
1 + d : C.INT	0.8796	0.9902	1	0.759	0.9394	1

2030			2050		
A	B1	B2	A	B1	B2
1.1504	1.7221	1.9544	1.1504	1.8914	2.4614
2.1927	4.7548	4.9775	2.6755	8.101	8.903
0.581	0.579	0.6619	0.523	0.521	0.5956
0.6071	0.7994	0.8619	0.3721	0.6228	0.7727

CASE III. POPULATION PRORATED APPORTIONING

$$DX = [(1+a)(1+b)(1+c)(1+d) \times GDP(t_o)i$$
$$\times E(t_o)_i \times C(t_o)_i$$

$$- \frac{\Sigma X(t_o)_i}{\Sigma P(t_o)_i}] \times P(t_o)_i \qquad (Eq.\ 3.9)$$

Similarly the amount of future CO_2 emission sacrifice is determined by the difference between current carbon output per capita multiplied by the growth potential factors and current world average per capita CO_2 emission.

Here, suppose we decided to stabilize global CO_2 emission at 1988 level, based on the above three criteria.

The difference of each alternative allowing emission quota and Bau emission is summarized in table 3.3-3.5. Developed countries obviously would be benefited from GDP or Present Level based apportioning, but even in the case of population prorated apportioning developed countries would face the least impact on their future energy supply once they pay a drastic short term restructuring cost. Even in the case of the most favorable case population prorated apportioning developing countries surprisingly might become the biggest loser eventually due to their strong population growth and relatively inferior energy and carbon intensity.

Table 3.3 Comparison of estimated sacrifice in CO_2 emission: developed countries (group A)

(unit:per capita C-Ton/yr)

TIME(YR)	Alt.I	Alt.II	Alt.III	PCT.BAUa	Sacrifice.I	Sacrifice.II	Sacrifice.III
1988	3.3099	3.7211	1.1360	3.3099	0.0000	0.4112	−2.1739
2000	3.1288	3.5175	1.0738	3.2981	−0.1692	0.2195	−2.2242
2010	3.0425	3.4205	1.0442	2.9156	0.1269	0.5049	−1.8714
2020	2.9587	3.3262	1.0154	2.5741	0.3846	0.7521	−1.5587
2030	2.8771	3.2345	0.9874	2.5607	0.3164	0.6738	−1.5733
2040	2.8771	3.2345	0.9874	2.1672	0.7099	1.0673	−1.1797
2050	2.8771	3.2345	0.9874	1.7237	1.1533	1.5108	−0.7363
2060	2.8771	3.2345	0.9874	1.6862	1.1909	1.5483	−0.6987
2070	2.8771	3.2345	0.9874	1.6442	1.2329	1.5904	−0.6567
2080	2.8771	3.2345	0.9874	1.6069	1.2702	1.6276	−0.6195
2090	2.8771	3.2345	0.9874	1.5654	1.3117	1.6691	−0.5780
2100	2.8771	3.2345	0.9874	1.5285	1.3486	1.7060	−0.5410

Table 3.4 Comparison of estimated sacrifice in CO_2 emission: Nies (Group B1)

(unit:per capita C-Ton/yr)

TIME(YR)	Alt.I	Alt.II	Alt.III	PCT.BAUb1	Sacrifice.I	Sacrifice.II	Sacrifice.III
1988	0.8567	0.8859	1.1360	0.8567	0.000	0.0293	0.2793
2000	0.7031	0.7271	0.9323	1.3631	−0.6600	−0.6359	−0.4307
2010	0.6265	0.6479	0.8308	1.6891	−1.0626	−1.0412	−0.8583
2020	0.5583	0.5773	0.7403	1.8501	−1.2918	−1.2728	−1.1098
2030	0.4975	0.5144	0.6596	1.8855	−1.3880	−1.3711	−1.2258
2040	0.4747	0.4909	0.6294	2.0798	−1.6051	−1.5889	−1.4503
2050	0.4529	0.4684	0.6006	2.2524	−1.7995	−1.7840	−1.6518
2060	0.4404	0.4555	0.5840	1.9063	−1.4658	−1.4508	−1.3222
2070	0.4283	0.4429	0.5679	1.5162	−1.0879	−1.0733	−0.9483
2080	0.4283	0.4429	0.5679	1.4832	−1.0549	−1.0403	−0.9153
2090	0.4283	0.4429	0.5679	1.4462	−1.0179	−1.0033	−0.8783
2100	0.4283	0.4429	0.5679	1.4135	−0.9852	−0.9705	−0.8455

Table 3.5 Comparison of estimated sacrifice in CO_2 emission: Other developing countries (Group B1)

(unit:per capita C-Ton/yr)

TIME(YR)	Alt.I	Alt.II	Alt.III	PCT.BAUb2	Sacrifice.I	Sacrifice.II	Sacrifice.III
1988	0.4205	0.2783	1.1360	0.4205	0.0000	−0.1422	0.7155
2000	0.3304	0.2187	0.8926	0.5708	−0.2404	−0.3521	0.3218
2010	0.2864	0.1895	0.7737	0.7024	−0.4160	−0.5128	0.0713
2020	0.2482	0.1643	0.6706	0.9195	−0.6713	−0.7552	−0.2489
2030	0.2151	0.1424	0.5813	1.1938	−0.9786	−1.0514	−0.6125
2040	0.1917	0.1269	0.5179	1.4383	−1.2466	−1.3114	−0.9203
2050	0.1708	0.1131	0.4615	1.7229	−1.5521	−1.6098	−1.2614
2060	0.1700	0.1125	0.4594	1.9712	−1.8012	−1.8587	−1.5119
2070	0.1692	0.1120	0.4572	2.2217	−2.0524	−2.1096	−1.7645
2080	0.1688	0.1117	0.4559	2.0419	−1.8731	−1.9302	−1.5859
2090	0.1683	0.1114	0.4547	1.7233	−1.5550	−1.6119	−1.2686
2100	0.1683	0.1114	0.4547	1.3745	−1.2062	−1.2631	−0.9198

(3) ELIGIBLE STABILIZATION PATH

It is quite certain from Figure 3 that any attempt to stabilize global CO_2 emission at 1988 level or ever below look like impractical. It is also clear that any one of the discussed allocation method cannot be acceptable to certain countries. There is a serious conflict of interests. Thus, the pathway to achieve stabilization of CO_2 emission should be based on the ultimate equity but flexible and self-designed path reflecting each individual country's particular situation.

Figure 5 represents the suggested path to achieve stabilization of CO_2 emission level approximately equal to the 2/3 of 1988 level until the end of 21st century, conceptually equal to the natural absorption level.

Here, natural absorption rate of CO_2 was assumed to be increased by 2bil.Ton Carbon per year through reforestation. It was estimated that 539.5 mil.ha of temperate zone forestry equivalent reforestation was necessary to build up the natural absorption rate by 2bil Ton Carbon per year, where the unit absorption capacity was assumed as 1.5 Ton Carbon per acre per year (Trexler, 1991). It means global forest area should be increased by 13.32 per cent comparing with 1988's 4.049 bil.ha.

Basically all the countries should reach same per capita CO_2 emission ultimately, but has different time table to get there according to the each country's development stage.

Developed countries would reduce their per capita CO_2 emission to the 1988 world average per capita until 2030, and further reduce to the 1/3 of 1988 level until 2050, and thereafter maintain the total emission.

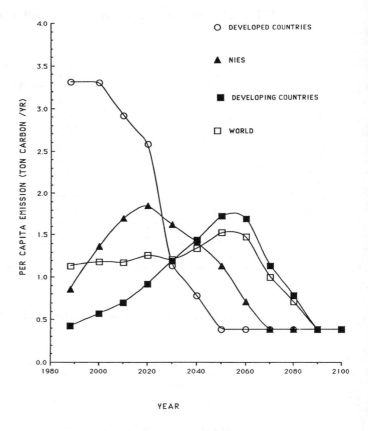

Figure 5. Suggested stabilization path

Developing countries would reduce their per capita CO_2 emission to the 1988 world average per capita emission until 2070, and further reduce to the 1/3 of 1988 level until 2090. Among developing countries, some countries like NIES might reduce their per capita emission to the 1988 world average level until 2050, and further reduce to the 1/3 of 1988 level until 2070, and similarly keep the total emission thereafter.

Table 3.6-1. Comparison of per capita CO_2 emission between suggested path and previous three alternatives : Group A

(unit:Ton Carbon/yr)

YEAR	POP.A* (mil.)	GDP.A* (U$'85)	EFF.A	MIX.A	CTT.A+	STB.88	GDP.88	PCT.88	PCT.STBa
1988	1224.8	10362	0.4181	0.764	4053986.	3.3099	3.7211	1.1360	3.3099
2000	1.0579	1.3767	0.3441	0.6720	4273268	3.1288	3.5175	1.0738	3.2981
2010	1.0879	1.7970	0.2700	0.5799	3884905	3.0425	3.4205	1.0442	2.9156
2020	1.1187	1.9850	0.2565	0.4879	3527067	2.9587	3.3262	1.0154	2.5741
2030	1.1504	2.1927	0.2430	0.2058	1600698	2.8771	3.2345	0.9874	1.1360
2040	1.1504	2.4221	0.2309	0.1341	1094810	2.8771	3.2345	0.9874	0.7770
2050	1.1504	2.6755	0.2187	0.0625	533565.5	2.8771	3.2345	0.9874	0.3787
2060	1.1504	2.8123	0.2078	0.0625	533565.5	2.8771	3.2345	0.9874	0.3787
2070	1.1504	2.9561	0.1968	0.0628	533565.5	2.8771	3.2345	0.9874	0.3787
2080	1.1504	3.1073	0.1870	0.0629	533565.5	2.8771	3.2345	0.9874	0.3787
2090	1.1504	3.2662	0.1771	0.0632	533565.5	2.8771	3.2345	0.9874	0.3787
2100	1.1504	3.4333	0.1683	0.0632	533565.5	2.8771	3.2345	0.9874	0.3787

* : ratio w.r.t. 1988 level
\+ : Total CO_2 Emission Of Group A(10^3 Ton Carbon)

Table 3.6-2. Comparison of per capita CO_2 emission between suggested path and previous three alternatives : Group B1

(unit:Ton Carbon/yr)

YEAR	POP.B1* (mil.)	GDP.B1* (U$'85)	EFF.B1	MIX.B1	CTT.B1	STB.88	GDP.88	PCT.88	PCT.STBb1
1988	298.6	2467	0.4663	0.7447	255803	0.8567	0.8859	1.1360	0.8567
2000	1.2184	1.7959	0.4171	0.7374	495913	0.7031	0.7271	0.9323	1.3631
2010	1.3674	2.6583	0.3682	0.6996	689649	0.6265	0.6479	0.8308	1.6891
2020	1.5345	3.5552	0.3191	0.6611	847745	0.5583	0.5773	0.7403	1.8501
2030	1.7221	4.7548	0.2700	0.5093	829488	0.4975	0.5144	0.6596	1.6131
2040	1.8048	6.2064	0.2565	0.3575	756717	0.4747	0.4909	0.6293	1.4042
2050	1.8914	8.1010	0.2430	0.2058	641596	0.4529	0.4684	0.6006	1.1360
2060	1.9451	8.9486	0.2309	0.1384	409590	0.4404	0.4555	0.5840	0.7052
2070	2.0002	9.8848	0.2187	0.0710	226166	0.4283	0.4429	0.5679	0.3787
2080	2.0002	10.3903	0.2078	0.0711	226166	0.4283	0.4429	0.5679	0.3787
2090	2.0002	10.9217	0.1968	0.0714	226166	0.4283	0.4429	0.5679	0.3787
2100	2.0002	11.4802	0.1870	0.0715	226166	0.4283	0.4429	0.5679	0.3787

* : ratio w.r.t. 1988 level

Table 3.6-3. Comparison of per capita CO_2 emission between suggested path and previous three alternatives : Group B2

(unit:Ton Carbon/yr)

YEAR	POP.B2* (mil.)	GDP.B2* (U$'85)	EFF.B2	MIX.B2	CTT.B2+	STB.88	GDP.88	PCT.88	PCT.STBb2
1988	3604.6	775.	0.6341	0.8556	1515609	0.4205	0.2783	1.1360	0.4205
2000	1.2727	1.5645	0.5502	0.8556	2618503	0.3304	0.2187	0.8926	0.5708
2010	1.4683	2.2717	0.4663	0.8556	3717610	0.2864	0.1895	0.7737	0.7024
2020	1.6940	3.3626	0.4430	0.7965	5614721	0.2482	0.1643	0.6706	0.9195
2030	1.9544	4.9775	0.4197	0.7374	8409889	0.2151	0.1424	0.5813	1.1938
2040	2.1933	6.6569	0.3987	0.6993	11370902	0.1917	0.1269	0.5179	1.4383
2050	2.4614	8.9030	0.3777	0.6611	15286224	0.1708	0.1131	0.4615	1.7229
2060	2.4730	11.9069	0.3588	0.5093	15033009	0.1700	0.1125	0.4594	1.6864
2070	2.4846	15.9244	0.3399	0.2708	10174096	0.1692	0.1120	0.4572	1.1360
2080	2.4916	17.5904	0.3229	0.1765	6978148	0.1688	0.1117	0.4559	0.7770
2090	2.4986	19.4308	0.3059	0.0822	3410383	0.1683	0.1114	0.4547	0.3787
2100	2.4986	21.4637	0.2906	0.0783	3410838	0.1683	0.1114	0.4547	0.3787

* : ratio w.r.t. 1988 level

Table 3.6-4. World CO_2 emission resulted from suggested path

YEAR	POP.W* (mil.)	GDP.W* (U$'85)	EFF.W	MIX.W	CTT.W+	++WORLD ACC.CO2	PCT.STBb2
1988	5128	3163	0.4575	0.7849	5825399		1.1360
2000	1.2182	4107	0.3933	0.7320	7387684	85.8850	1.1826
2010	1.3716	5233	0.3268	0.6893	8292164	164.7365	1.1790
2020	1.5473	6064	0.3234	0.6420	9989534	256.9936	1.2590
2030	1.7488	7273	0.3191	0.5208	10840075	361.5670	1.2087
2040	1.9216	8565	0.3145	0.4982	13222430	483.0707	1.3418
2050	2.1151	10286	0.3084	0.4784	16461385	633.1092	1.5177
2060	2.1264	12486	0.3012	0.3896	15976164	795.0543	1.4652
2070	2.1378	15349	0.2927	0.2220	10933827	927.0831	0.9974
2080	2.1426	16666	0.2796	0.1511	7737879	1018.843	0.7042
2090	2.1476	18107	0.2663	0.0785	4170115	1076.599	0.3787
2100	2.1476	19692	0.2543	0.0756	4170115	1118.300	0.3787

* : ratio w.r.t. 1988 level

+ : Annual CO_2 Emission (Thousand Ton Carbon)

++ : Accumulated CO_2 Emission (Bil. Ton Carbon)

According to the derived relationship between the accumulated man-made CO_2 emission and atmospheric CO_2 concentration, the future carbon emission budget to allow atmospheric CO_2 concentration to reach a certain level, or the atmospheric CO_2 concentration resulted from the proposed path can be estimated.

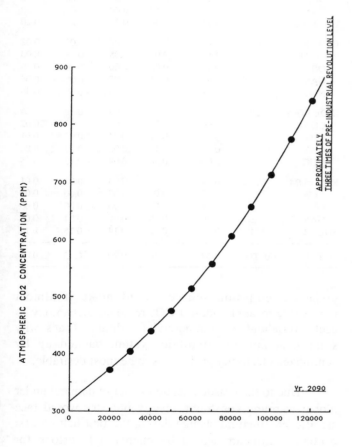

Figure 6. Projection of CO_2 concentration as a function of accumulated emission

Adopting this path, it may be expected that the global CO_2 emission decreases to approximately 2/3 of 1988 level until the end of next century and atmospheric CO_2 concentration would reach approximately three times of pre-Industrial Revolution level (3 × 278 ppm). More importantly, no nation would not be hurt unfairly in terms of their economic growth.

Needless to say actual time table for this stabilization path could be either accelerated or delayed by new facts based on the best knowledge that science can provide.

IV. DISCUSSION

It looks almost impossible to stabilize CO_2 emission approximately at a level of 1988's 6 bil.Ton-Carbon per year within the near future. Only the developed countries would be able to lower the total CO_2 emission down to 1988's level until the middle of 21st century.

Developing countries, mainly due to their economic growth, cannot but increase CO_2 emission more than 5 times then 1988 level until 2070. This is nothing but a reflection of today's world which majority of mankind just marginally survive on less than a 1/10 of developed countries' per capita income.

Table 4.1 shows the 21st century wealth distribution prospect which global CO_2 emission is based on. It implies much more improved economic parity between developed countries and developing countries.

As shown in Table 4.1 economic growth rate of NIES and other developing countries was assumed 2.37 per cent and 2.95 per cent respectively for the next century. These economic growth assumptions moderately reflect growth target planned by most of developing countries' government. At the same time it should be recognized that CO_2 emission stabilization based on current output to curb total emission at 1988 level would limit developing countries' future economic growth, far below the level otherwise achieved, up to 1.16 per cent for NIES and 0.63 per cent for the other developing countries, while developed countries might lose nothing.

Table 4.1. Comparison of per capita GDP by country group

(unit : U$ '85 Const.)

Year	Sugg. Path			1988 Freezing		
	A	B1	B2	A	B1	B2
1988	10362 (1.0)	2467 (0.24)	775 (0.07)	10362	2467	775
2010	18621	6558	1649	18621	2432	718
2030	22721	11730	3858	22721	3095	695
2050	27724	19985	6900	27724	4019	684
2070	30631	24386	12341	30631	6888	940
2090	33844 (1.0)	26944 (0.80)	15059 (0.45)	33844 (1.0)	7979 (0.24)	1471 (0.04)
G.Rate (%)	1.17	2.37	2.95	1.17	1.16	0.63

( ) : fraction with respect to Group A Per Capita GDP

Provided that population growth could reach a standstill until the end of 21st century and economic growth would continue to achieve better living standard, the possible way to reduce the total CO_2 emission would be to lower the carbon intensity in energy mix and to enhance the overall efficiency of energy.

In relation to the overall energy efficiency factor, it should be recognized that even though industrial structure change toward less energy intensive and high value added industry must be critical to improve an individual country's overall energy efficiency in terms of consumed energy per unit GDP, it could not contribute to reduce the global CO_2 emission as long as the production technology itself remains same. In other words if an energy intensive industry moves from a developed country to a developing country without improving the technology, it would be a just redistribution rather than a real reduction in CO_2 emission.

At the moment energy conservation has quite good potential to decrease total energy demand. Probably energy conservation must have a certain limit imposed by physical law and be saturated as time goes by. That's why material recycling is crucial to minimize total energy demand of a society and consequently to enhance the overall energy efficiency. Especially the energy intensive raw materials such as iron, non-ferrous metals, glass, paper, and plastics should be recycled as much as possible.

Table 4.2 is the illustration of energy supply option by demand sector to meet the required CO_2 emission cut. As shown in Table 4.2 a drastic change in energy supply for next couple of decades is inevitable to stabilize the CO_2 emission until the end of 21 st century.

More specifically, in transportation sector present petroleum powered car should be replaced most desirably by Hydrogen Fuel Cell powered car. In power generating sector, nuclear energy looks like the only practical solution in large scale centralized power system. Gas fired power station might play its role as a peak shaver. It seems that CO_2 removal process should be added eventually to the coal fired power station in order to meet the CO_2 reduction target. Otherwise coal would cease to be a fuel for power generation. Even in the industry sector nuclear generated electricity and heat would be required to replace most of fossil fuel. Nevertheless, gas for direct heating, coal for coke making or direct reduction of iron ore would remain competitive over nuclear energy. Residential and commercial sector is composed of a huge number of small to mid-scale consumers. This characteristics of

Table 4.2. Carbon intensity illustration of developed countries

SECTOR	OECD.88	2030 (Bau)	2030	2040	2050
TRANSP	0.20	0.16	0.14	0.00	0.00	0.00
OIL		0.20	0.10	0.00	0.00	0.00
GAS		0.00	0.10	0.00	0.00	0.00
ELC.HYD		0.00	0.00	0.20	0.20	0.20
POWER	0.40	0.22	0.08	0.08	0.06	0.02
COAL		0.17	0.05	0.05	0.03	0.00
OIL		0.03	0.00	0.00	0.00	0.00
GAS		0.03	0.05	0.05	0.05	0.04
NUC		0.17	0.30	0.30	0.32	0.36
INDUSTRY	0.20	0.16	0.15	0.09	0.06	0.04
COAL		0.06	0.04	0.04	0.03	0.02
OIL		0.07	0.06	0.04	0.02	0.01
GAS		0.06	0.10	0.03	0.03	0.02
NUC		0.00	0.00	0.09	0.12	0.15
RES. COM	0.20	0.14	0.10	0.03	0.02	0.01
COAL		0.02	0.01	0.00	0.00	0.00
OIL		0.08	0.04	0.00	0.00	0.00
GAS		0.09	0.10	0.06	0.04	0.01
SOLAR	0.01	0.05	0.05	0.14	0.16	0.19
TOTAL	1.0 TOE	0.68	0.46	0.20	0.15	0.07

dispersed emission source would make it almost impossible to apply add-on CO_2 removal technology to each household or commercial building. That's why solar heat and photovoltaic system backed up by centralized electricity grid seems to be most desirable.

Due to the drastic change in energy demand under global warming constraint, energy industry would face significant structural change during the first half of 21st century. Petroleum would be mainly a feedstock for petrochemical industry rather than a mere fuel to burn out. If coal were replaced by gas, per calorie CO_2 emission could be reduced by 44 per cent. Even though gas could reduce CO_2 emission considerably, expanded gas usage would be an intermadiate step to mitigate CO_2 emission rather than an ultimate alternative. High economic growth potential in developing countries and already high energy consumption would easily offset the improved effect by fuel switching in terms of total emission. If we switched other fuels to gas by 100 per cent, the aggregate emission factor would be 0.577 per cent Ton C/TOE. This must be significantly lower than current level (World Average as of 1988:0.7849).

It can be observed from Table 4.2 that it would fail to meet the required carbon intensity from the year around 2030 with expanded gas usage alone. Coal industry looks like the biggest loser under new environmental constraint. If it were possible to develop effective add-on CO_2 removal process, coal could remain competitive in power and industry sector composed of large scale consumers over any non-conventional energy.

V. CONCLUSION AND RECOMMENDATION

1. **Atmospheric CO_2 concentration vs. accumulated man-made CO_2 emission data are excellently fitted by exponential regression.**

2. **Developing Countries Future Economic Growth**

Any allocation method to stabilize CO_2 emission at 1988 levels would hurt developing countries economic growth potential seriously.

    (1)   Current Emission Level Based Allocation Obviously it is unfavorable to developing countries.

    (2)   GDP based Allocation

        It is not so much different as Current Emission Level Based Allocation.

    (3)   Population Prorated Allocation

        For short term period, it is most favorable to developing countries and requires a drastic restructuring cost to developed countries. Surprisingly it is not so much favorable to developing countries as it is generally conceived to be in the long run. Even though this allows the largest CO_2 emission quota to developing countries, it is still not big enough to meet the developing countries' future energy demand to support the economic growth.

3. **Eligible Stabilization Path**

Under given population growth, a tentative goal of annual global carbon emission might be set up as circa 4 bil. Ton Carbon/yr at the and of 21st century. With this emission stabilization path the atmospheric CO_2 concentration might reach approximately 800 ppm, triple of the Pre-industrial Revolution level.

Eventual per capita emission equity should be applied until the end of 21 st century, but as long as accumulated emission equity were kept, each country should be allowed to establish a different time table to reach the ultimate equity point based on individual country's development stage.

4. **Energy Policy Implication**

    (1)   Energy conservation coupled with material recycling is most desirable to minimize total energy demand.

    (2)   Nuclear Power is inevitable to lower the carbon intensity within the near future.

    (3)   New and renewable energy R&D should be focused on non-CO_2 emitting and high potential application energy resources such as solar and hydrogen fuel cell derived from non-fossil fuel resources.

    (4) CO_2 sink control technology should be developed such as artificial photosynthesis or reforestation of fast growing trees.

    (5)   Energy industry would be more technology intensive than ever before due to high efficiency requirement, non-fossil fuel utilization and fossil fuel CO_2 control. Accordingly, technology gap between developed countries and developing countries would be widened and technology dependency of developing countries on developed countries would be worsened.

    (6)   Energy pricing and tax system would incorporate environmental externality. Incentives for new and renewable energy promotion and carbon tax to cope with global warming look inevitable.

RECOMMENDATION

1. **In relation to international carbon sink management,**

Global carbon fund should be established and operated as matching fund with domestic carbon fund. To expedite the establishment of global carbon fund, it is necessary to start with a very low rate, e.g., U$ 0.1 per TOE. Each country shall be allowed to be subject to different effective date based on individual country's development stage, e.g., 2000 for OECD countries and mid-21st century for developing countries.

If necessary, CO_2 removal liability allocation should be introduced based on historical contribution share and technological level.

2. In relation to long term energy policy to lower carbon intensity,

Domestic carbon fund collected from fossil fuels in desirable to promote energy conservation and new and renewable energy resources development.

Material recycling system such as Bottle deposit law is highly desirable to enhance the recycle ratio of energy intensive raw materials.

Public acceptance is the bottle neck for nuclear option. Advance site acquisition or reservation for nuclear power plants is highly desirable even in those developing countries not currently taking nuclear option to broaden future energy supply option under global warming constraint.

REFERENCES

British Petroleum, BP Statistical Review of World Energy, London: Corporate Communications Services, 1991:34

International Energy Agency, Energy Balances of OECD Countries 1987-1988, Paris: OECD 1990:70

International Energy Agency, Energy Balances of OECD Countries 1987-1988, Paris: OECD 1990:121

International Energy Agency, Greenhouses Gas Emissions- The Energy Dimension, Paris: OECD 1991:27

International Monetary Fund, International Financial Statistics, Washington D.C.: IMF, 1989:445

Lapedes, D. N., McGraw-Hill Encyclopedia of Science and Technology New York: McGraw-Hill, 1977:679

Trexler, M.C., Minding the Carbon Store, Washington D.C.: World Resources Institute, 1991:25

United Nations Environment Programme, Environmental Data Report, 2nd. ed., Oxford: Basil Blackwall Inc. 1989:17

United Nations Environment Programme, Environmental Data Report, 2nd. ed., Oxford: Basil Blackwall Inc. 1989:21

United States DOE Multi-Laboratory Climate Change Committee, Energy and Climate Change, Chelsea Michigan : Lewis Publisher Inc. 1990:34